Handbook of Statistics

NIPA® GENX ELECTRONIC RESOURCES & SOLUTIONS P. LTD.
New Delhi-110 034

About the Authors

Dr. Debasish Biswas is presently serving as an Associate Professor in the Department of Business Administration, at Vidyasagar University, West Bengal. He is having Post – Graduate teaching experience of more than 14 years. His areas of specialization are Accounting & Finance, Human Resources, and Marketing. He has presented 105 research papers in different areas of management at different National and International Seminars and Conferences. He has authored 84 research papers in National and International journals of repute. He is a life member of the Indian Commerce Association and All India Accounting Association. He has authored twenty-six books. He has also contributed 21 chapters in various edited books. He received the Best Business Academic of the Year Award -2016 (Gold Medal) at the 69th All India Commerce Conference organized by Lucknow University, Lucknow. He received Bharat Vikas Award in 2018 from the Institute of Self Reliance, Odisha. He has successfully guided 10 Ph.D. Scholars.

Atanu Manna is a distinguished senior research scholar at the department of Centre for Environmental Studies at Vidyasagar University. Holding an MSc in Geography and an MBA with HR specialization, he has successfully qualified for UGC NET and JRF in Geography, and NET in HRM/Labour Welfare. He has an impressive academic portfolio, including the publication of nine research papers, two book chapters, and a book, showcasing his proficiency in blending academic research with practical data analysis. With two years of experience as a data analyst in the corporate sector, his work is a testament to making complex concepts accessible and applicable in the real world.

Dr. Tanusree Chakraborty holds a Ph.D. in Applied Psychology from Calcutta University, complemented by an MA in Applied Psychology with a specialization in Organizational Psychology, a PGDHRM, and an MBA in HR. With over 20 years of professional experience, she has contributed significantly to the academic and research spheres. Her association with esteemed B-Schools and Universities in Kolkata, Madurai, Bangalore, and Chennai highlights her commitment to academic excellence. Currently serving as a Faculty in the Centre for Management Studies at the Administrative Staff College of India (ASCI), Hyderabad, Dr. Chakraborty brings a wealth of expertise to her role. Dr. Chakraborty has authored approximately 70 publications in esteemed journals and books. She has delivered over 80 research papers at international conferences. Dr. Chakraborty has published books with IGI Global Publishers, Taylor and Francis, and CRC Press

Handbook of Statistics

Debasish Biswas
Atanu Manna
Tanusree Chakraborty

NIPA® GENX ELECTRONIC RESOURCES & SOLUTIONS P. LTD.
New Delhi-110 034

NIPA® GENX ELECTRONIC RESOURCES & SOLUTIONS P. LTD.

101,103, Vikas Surya Plaza, CU Block
L.S.C. Market, Pitam Pura, New Delhi-110 034
Ph : +91-11-43860225, Mob.: +91 9717133558, 9540816132
E-mail: newindiapublishingagency@gmail.com
Website: www.nipaersources.com

Print ISBN: 978-93-58874-88-4
ebook ISBN: 978-93-58876-70-3

NIPA® also publishes books in a variety of electronic formats. Some content that appears in print may not be available in electronic books, and vice versa.

Composed and Designed by NIPA®.

Preface

Welcome to the Handbook of Statistics for Professional Courses. This comprehensive guide aims to provide students pursuing professional courses in various fields with a solid foundation in statistics. Statistics plays a crucial role in decision-making and problem-solving in today's data-driven world, and this handbook is designed to equip readers with the knowledge and skills needed to understand and apply statistical concepts effectively.

The primary aim of this handbook is to explain statistics and make it accessible to students from diverse educational backgrounds. By presenting complex statistical concepts clearly and concisely, this book aims to help students develop a strong understanding of the subject. The objectives of this handbook include introducing the fundamental concepts of statistics, explaining the different types of statistics and variables, providing a comprehensive overview of data collection, presentation, and analysis techniques, equipping students with the skills needed to interpret and analyze data effectively, and enhancing students' understanding of correlation, regression, index numbers, sampling techniques, hypothesis testing, and probability distributions.

This handbook is tailored to meet the needs of students in professional fields such as business, economics, management, and social sciences, among others. These disciplines often require a strong foundation in statistics for data analysis, decision-making, and research purposes. Moreover, professionals working in these fields can also benefit from this handbook, as it provides a practical and comprehensive overview of statistical concepts that are directly applicable to their work. Whether you are a student looking to enhance your understanding of statistics or a professional seeking to refresh your statistical knowledge, this handbook is designed to meet your specific needs and help you excel in your academic or professional endeavours.

The Handbook of Statistics for Professional Courses offers clear and concise explanations of complex statistical concepts, practical examples, and case studies to illustrate key concepts, step-by-step guidance on data collection, analysis, and interpretation, comprehensive coverage of various statistical techniques and methods, and an emphasis on real-world applications of statistics in professional settings. What sets this handbook apart is its focus on the specific needs of students studying professional courses. The content is tailored to provide practical knowledge and skills that are directly applicable to their academic and professional pursuits. Additionally, the inclusion of real-world examples and case studies enhances the relevance and practicality of the material presented.

This book is a valuable resource for students and professionals alike. Whether you are new to statistics or looking to enhance your existing knowledge, this handbook will serve as a comprehensive guide to understanding and applying statistical concepts in a professional context. We hope you find this handbook informative and engaging, and we wish you success in your statistical endeavours.

14th March, 2024
Midnapore

D. Biswas
A. Manna
T. Chakraborty

Contents

Preface v

Unit-I: Introduction To Statistics 1

1.1 Learning Objectives 1
1.2 Introduction 1
1.3 Meaning and Definition of Statistics 1
1.4 Functions of Statistics 3
1.5 Scope of Statistics 4
1.6 Limitations of Statistics 5
1.7 Types of Statistics 6
1.8 Statistics and Parameter 8
1.9 Types of Variables 9
1.10 Role of Various Statistical Organizations 17
1.11 Summary 19
1.12 Self-Assessment 19

Unit-II: Scale of Measurement and Data Structure 21

2.1 Learning Objectives 21
2.2 Introduction to Measurement in Statistics 21
2.3 Measurement's Function in Statistics 22
2.4 The Primary Functions of Measurement in Research 23
2.5 Uses of Measurement 24
2.6 Scales of Measurement 26
2.7 Types of Scales 28
2.8 Significance of Different Scales 38
2.9 Some Important Scales 39
2.10 Data Structure 41
2.11 Future Trends in Data Structure Analysis 49
2.12 Recent Data Structure Trends 50
2.13 Summary 51
2.14 Self-Assessment 52

Unit-III: Collection and Presentation of Data 53
3.1 Learning Objectives 53
3.2 Introduction 53
3.3 Importance of Data Collection and Presentation 53
3.4 Role of Data in Decision-Making 54
3.5 Primary Data and Its Collection Techniques 54
3.6 Secondary Data and Its Collection Techniques 56
3.7 Frequency Distribution 58
3.8 Relative Frequency Distribution 61
3.9 Cumulative Frequency Distribution 62
3.10 Presentation of Data 63
3.11 Chapter Summary 77
3.12 Self-Assessment 77

Unit-IV: Measures of Central Tendency and Partition Values 79
4.1 Learning Objectives 79
4.2 Introduction 79
4.3 Measures of Central Tendency 79
4.4 Partition Values-Quartiles, Deciles, and Percentiles 102
4.5 Chapter Summary 110
4.6 Self-Assessment 110

Unit-V: Measures of Dispersion 111
5.1 Learning Objectives 111
5.2 Introduction 111
5.3 Measures of Dispersion 111
5.4 Shape of Distribution 131
5.5 Chapter Summary 139
5.6 Self-Assessment 139

Unit-VI: Correlation 141
6.1 Learning Objectives 141
6.2 Introduction 141
6.3 Meaning of Correlation 141
6.4 Significance of Correlation 143
6.5 Correlation Coefficient and Causation 144

6.6 Various Kinds of Correlation....................145
6.7 Properties of Correlation Coefficient....................147
6.8 Scatter Diagram....................149
6.9 Karl Pearson's Correlation Coefficient....................154
6.10 Spearman's Rank Correlation Coefficient....................157
6.11 Standard and Probable Error of Correlation....................160
6.12 Coefficient of Determination (R^2)....................160
6.13 Partial Correlation....................161
6.14 Multiple Correlation....................163
6.15 Limitations of Correlation....................167
6.16 Summary....................168
6.17 Self-Assessment....................169

Unit-VII: Regression 171

7.1 Learning Objectives....................171
7.2 Introduction....................171
7.3 Meaning of Regression....................172
7.4 Significance of Regression....................177
7.5 Regression vs. Correlation....................178
7.6 Regression Lines....................179
7.7 Properties of Regression Coefficient....................181
7.8 Simple and Multiple Regressions....................182
7.9 Concept of R Square in Regression....................191
7.10 Summary....................191
7.11 Self-Assessment....................192

Unit -VIII: Index Numbers....................195

8.1 Learning Objectives....................195
8.2 Introduction....................195
8.3 Definition and Overview of Index Number....................195
8.4 Importance of Index Number in Statistics....................196
8.5 Historical Background....................196
8.6 Base Period and Current Period....................196
8.7 Methods of Constructing Index Numbers....................197
8.8 Chain Base Index Numbers....................209
8.9 Quantity Index Number....................211

8.10 Properties of Good Index Numbers 211
8.11 Different Types of Index Numbers 216
8.12 Purchasing Power of Money, Deflating Using Index Number and Real Wage 225
8.13 Base Shifting and Splicing of Index Numbers 227
8.14 Practical Applications of Index Numbers 229
8.15 Challenges in Using Index Numbers 230
8.16 Recent Advances and Future Trends 230
8.17 Summary 231
8.18 Self-Assessment 232

Unit-IX: Sampling Technique 235

9.1 Learning Objectives 235
9.2 Introduction 235
9.3 Significance of Sampling in Research and Data Analysis 239
9.4 Sampling Frame 240
9.5 Sample Size 246
9.6 Sampling Techniques 253
9.7 Sampling and Non-Sampling Errors 274
9.8 Conclusion 276
9.9 Summary 277
9.10 Self-Assessment 277

Unit-X: Hypothesis Testing 279

10.1 Learning Objectives 279
10.2 Introduction 279
10.3 Meaning of Hypothesis Testing 279
10.4 Importance of Hypothesis Testing in Scientific Inquiry and Decision-Making 282
10.5 Types of Hypotheses 283
10.6 One-Tailed & Two-Tailed Tests 287
10.7 Types of Error 288
10.8 Degrees of Freedom 289
10.9 Confidence Limits 290
10.10 Steps in Hypotheses Testing 291

10.11 Summary 300
10.12 Self-Assessment 300

Unit-XI: Probability Distribution 303

11.1 Learning Objectives 303
11.2 Introduction 303
11.3 Meaning of Probability Distribution 304
11.4 Binomial Distribution 305
11.5 Poisson Distribution 310
11.6 Normal Distribution 313
11.7 Summary 319
11.8 Self-Assessment 320

***Appendix*..323**

***Bibliography*..329**

***Index*..331**

Unit-I

Introduction to Statistics

1.1 Learning Objectives

After going through this chapter, you will be able to:

i) Understand the meaning, scope, functions, and limitations of statistics;

ii) Know the various types of statistics and their importance;

iii) Comprehend the various types of variables;

iv) Appreciate the role of the various statistical organizations.

1.2 Introduction

Statistics is a powerful and essential field of study with substantial importance in various aspects of life, science, business, and society. Its significance lies in its ability to provide a systematic framework for collecting, analyzing, interpreting, and presenting data. Statistics is a universal language of data that empowers individuals, organizations, and societies to make informed decisions, solve problems, and gain deeper insights into the world around us.

1.3 Meaning and Definition of Statistics

Statistics is a multidisciplinary field that involves the collection, analysis, interpretation, presentation, and organization of data. It encompasses a wide range of techniques and methods used to summarize and make sense of numerical and categorical information. Statistics is employed in various fields, such as science, business, economics, social sciences, government, and many others, to draw conclusions, make informed decisions, and understand patterns and trends within data.

Statistics is a branch of applied mathematics that involves the collection, description, analysis, and inference of conclusions from quantitative data.

Statistics may be defined as the collection, presentation, analysis, and interpretation of numerical data. It consists of the following steps.

1. Collection of Data

In statistics, the collection of data refers to the process of systematically gathering information or observations about individuals, objects, events, or phenomena of interest. Data collection is a crucial step in the statistical analysis process because the quality and reliability of the collected data significantly impact the validity of any subsequent analysis or interpretation. Accurate and reliable data are essential for making informed decisions, drawing meaningful conclusions, and achieving research objectives in various fields, including science, business, healthcare, social sciences, etc.

2. Organisation of Data

The organization of data refers to the process of structuring, arranging, and presenting data in a way that makes it understandable, interpretable, and appropriate for analysis. Organizing data involves transforming raw data into a structured format that simplifies data exploration, summary, and analysis. Properly organized data sets the stage for meaningful statistical analysis and the extraction of valuable insights.

3. Presentation of Data

The presentation of data in statistics refers to the various methods and techniques used to effectively communicate and convey information derived from data to the various stakeholders. Proper data presentation is crucial as it helps to make data more understandable, informative, and visually attractive. It allows individuals, researchers, analysts, and decision-makers to convey the meaning and insights behind the data to the various stakeholders. Usually, data are presented in the form of tables, diagrams, and graphs.

4. Analysis of Data

Analysis of data in statistics refers to the process of examining, exploring, and drawing meaningful conclusions from a dataset. It involves applying various statistical techniques, methods, and tools to uncover patterns, relationships, trends, and insights within the data. The primary goal of data analysis is to transform raw data into actionable information, answer research questions, make informed decisions, and support hypotheses or claims.

5. Interpretation of Data

Interpretation of data in statistics refers to the process of making sense of the results obtained from data analysis. It involves translating statistical findings, patterns, and relationships into meaningful insights, conclusions, and actionable recommendations. Data interpretation is a critical step in the data analysis process as it allows individuals, researchers, and decision-makers to draw meaningful inferences and understand the implications of the data.

1.4 Functions of Statistics

Statistics serves several important functions in various fields and applications. These functions are crucial for understanding data, making informed decisions, and gaining insights into complex phenomena. Here are some of the key functions of statistics:

i) **Description:** Statistics provides a means to summarize and describe data in a concise and meaningful way. It helps in presenting data through various measures like averages, percentages, and graphical representations, making it easier to understand and interpret.

ii) **Inference:** Statistics enables us to make inferences about a population based on a sample from that population. This function is essential for drawing conclusions and making predictions, especially when it is impractical or impossible to study an entire population.

iii) **Data Collection:** Statistics guides the process of data collection, helping to design surveys, experiments, and observational studies in a systematic and impartial manner. It ensures that data is collected efficiently and effectively.

iv) **Data Analysis:** Statistics provides tools and techniques for analyzing data, and identifying patterns, relationships, and trends within datasets. This analysis helps to uncover valuable insights and knowledge.

v) **Decision Making:** Statistics plays a critical role in decision-making processes. It offers a quantitative basis for making choices, evaluating options, and assessing risks. Businesses, governments, and individuals use statistical analysis to inform their decisions.

vi) **Quality Control:** Statistics is widely used in quality control and quality assurance processes in manufacturing and service industries. It helps to detect defects, monitor processes, and ensure product quality.

vii) **Research:** In scientific research, statistics helps to design experiments, analyze experimental results, and determine the significance of findings. It aids researchers in drawing valid conclusions from their studies.

viii) **Prediction and Forecasting:** Statistics is employed in forecasting future trends and outcomes. Methods like time series analysis and regression analysis are used to make predictions based on historical data.

ix) **Risk Assessment:** Statistics is involved in assessing and managing risks. It quantifies uncertainty and provides a framework for evaluating the likelihood and impact of various risks.

x) **Policy Development:** Governments and policymakers use statistical data and analysis to formulate policies, allocate resources, and address societal issues. It helps them to understand the needs of populations and measure the effectiveness of policies.

xi) **Benchmarking:** Organizations use statistical benchmarks to compare their performance with industry standards or competitors. This helps to identify areas for improvement and set performance goals.

xii) **Communication:** Statistics aids in the effective communication of data and information. Visual representations like charts, graphs, and tables make complex data more accessible and understandable to a wider audience.

Overall, statistics serves as a multipurpose and indispensable tool for understanding the world, making informed decisions, and driving progress in numerous fields, from science and business to public policy and healthcare.

1.5 Scope of Statistics

The scope of statistics is enormous and encompasses a wide range of applications across various fields and industries. As data becomes increasingly abundant and essential in the modern world, the scope of statistics continues to expand. Here are some key areas where statistics plays a significant role:

i) **Science and Research:** Statistics is fundamental to scientific research, including fields like biology, physics, psychology, and environmental science. It helps to design experiments, analyze data, and draw valid conclusions.

ii) **Business and Economics:** Statistics is extensively used in business for market research, financial analysis, quality control, and operations management. It informs decision-making and strategic planning.

iii) **Healthcare and Medicine:** Statistics is critical in clinical trials, epidemiology, and healthcare management. It aids in evaluating the effectiveness of treatments, studying disease patterns, and optimizing healthcare delivery.

iv) **Government and Public Policy:** Governments rely on statistics for census data, economic indicators, and policy development. It informs decisions related to resource allocation, infrastructure planning, and social programs.

v) **Education:** Educational institutions use statistics for assessment and evaluation purposes. It helps to analyze student performance, improve teaching methods, and assess educational policies.

vi) **Finance and Banking:** Financial institutions use statistics for risk assessment, portfolio management, and fraud detection. It plays a crucial role in investment decisions and financial modeling.

vii) **Social Sciences:** Sociology, psychology, political science, and other social sciences use statistics to analyze survey data, study human behaviour, and draw conclusions about populations.

viii) **Environmental Science:** Environmental scientists use statistics to analyze data related to climate change, pollution, biodiversity, and natural resource management.

ix) **Marketing and Market Research:** Statistics is essential for market segmentation, consumer behaviour analysis, and product development. It helps businesses to target their audience effectively.

x) **Sports Analytics:** Statistics plays a growing role in sports analytics, helping teams and athletes to optimize performance, make strategic decisions, and evaluate player statistics.

xi) **Quality Control and Manufacturing:** Industries rely on statistical process control to maintain quality standards and improve production processes.

xii) **Data Science and Artificial Intelligence:** Statistics is a foundational component of data science and machine learning. It provides the tools and techniques for data analysis, modeling, and prediction.

xiii) **Insurance:** The insurance industry uses statistics to assess risk and set premium rates. Actuaries rely on statistical models to make predictions about future events.

xiv) **Energy and Utilities:** Statistics is used for energy consumption analysis, grid management, and resource allocation in the energy and utilities sector.

xv) **Market Economics:** Economists use statistics to analyze economic trends, inflation, employment rates, and other macroeconomic factors.

The scope of statistics continues to expand as organizations increasingly recognize its importance in making data-driven decisions and solving complex problems.

1.6 Limitations of Statistics

The important limitations of statistics are as follows:

i) **Sampling Error:** When working with samples of data rather than entire populations, there is always the potential for sampling error. This means that the sample statistics may not perfectly represent the true population parameters. The accuracy of estimates depends on the size and representativeness of the sample.

ii) **Bias:** Bias can be introduced at various stages of data collection, analysis, or interpretation. Selection bias, measurement bias, and reporting bias are examples. Bias can lead to incorrect conclusions or skewed results.

iii) **Assumption Dependence:** Many statistical techniques rely on specific assumptions about the data distribution, independence of observations, or other factors. Violations of these assumptions can lead to inaccurate results.

iv) **Influence of Outliers:** Outliers, extreme values in the data, can have a disproportionate impact on statistical measures like the mean. They can skew results and misrepresent the central tendency of the data.

v) **Context-Dependence:** Statistical results may not always apply universally. Findings may be context-dependent and may not generalize to different populations, time periods, or settings.

vi) **Complexity:** Some statistical methods, especially advanced techniques, can be complex and require expertise to use effectively. Misapplication of complex statistical methods can lead to errors.

1.7 Types of Statistics

Statistics can be broadly categorized into two main types: descriptive statistics and inferential statistics. These types serve distinct purposes in the field of statistics and are used for different stages of data analysis and decision-making:

1. Descriptive Statistics

Descriptive statistics involve methods and techniques used to summarize, organize, and describe the main features of a dataset. They are primarily concerned with providing a clear and concise summary of the data, making it more understandable and interpretable. Descriptive statistics do not involve making inferences or predictions about a larger population; instead, they focus on characterizing the data at hand. Common descriptive statistics include Mean, Median, Mode, Range, Standard Deviation, Skewness, and Kurtosis. A very brief description is outlined here.

i) Measures of Central Tendency

- **Mean (Average):** The mean is the sum of all data values divided by the number of data points. It represents the center of the data distribution.
- **Median (Middle Value):** The median is the middle value when the data is sorted according to ascending order. It is less affected by extreme values and represents the middle most position in the arranged dataset.

- **Mode (Most Frequent Value):** The mode is the value that appears most frequently in the dataset. A dataset can have one mode (unimodal) or more than one mode (multimodal).

ii) Measures of Variability or Dispersion

- **Range:** The range is the difference between the maximum and minimum values in the dataset. It measures the spread or variability of the data.
- **Variance:** Variance measures how data points deviate from the mean. It is calculated as the average of the squared differences between each data point and the mean.
- **Standard Deviation:** The standard deviation is the square root of the variance. It provides a more interpretable measure of variability.

iii) Frequency Distributions

- **Frequency Distribution Table:** This table displays how often each value or range of values occurs in a dataset. It includes the frequency (count) and often the relative frequency (percentage) of each value or range.
- **Histogram:** A histogram is a graphical representation of the frequency distribution, showing data distribution in the form of bars. It provides a visual representation of data's shape and spread.
- **Frequency Polygon:** A frequency polygon is a line graph that represents the frequency distribution. It connects the midpoints of the bars in a histogram.

iv) Percentiles and Quartiles

- **Percentiles:** Percentiles divide the data into 100 equal parts. For example, the 25th percentile represents the value below which 25% of the data falls.
- **Quartiles:** Quartiles divide the data into four equal parts. The first quartile (Q1) represents the 25th percentile, the second quartile (Q2) is the median, and the third quartile (Q3) is the 75th percentile.

v) Summary Tables

Summary tables provide a compact representation of data, including counts, percentages, and summary statistics for different categories or groups within the dataset. They are useful for comparing and contrasting data subsets.

vi) Box-and-Whisker Plots (Boxplots)

Boxplots display the distribution of data by showing the median, quartiles, and potential outliers. They are useful for visualizing the spread and skewness of data.

vii) Measures of Skewness and Kurtosis

- **Skewness:** Skewness measures the asymmetry of the data distribution. Positive skew indicates a longer right tail, while negative skew indicates a longer left tail.
- **Kurtosis:** Kurtosis measures the 'tailedness' of the distribution. It assesses how peaked or flat the distribution is compared to a normal distribution.

Descriptive statistics provide an initial understanding of data, making it easier to identify patterns, trends, and key characteristics. These components help individuals to communicate and interpret data effectively, whether for research, decision-making, or reporting purposes.

2. Inferential Statistics

Inferential statistics involve methods and techniques used to make inferences or predictions about a larger population based on a sample of data. Inferential statistics allow researchers and analysts to draw conclusions, test hypotheses, and make decisions by extrapolating from the sample to the population.

1.8 Statistics and Parameter

Statistics and parameters are two fundamental concepts in statistics, and they have distinct meanings and roles in the analysis of data. Here's a more detailed explanation of each term:

i) Statistics

Definition: Statistics refer to numerical values calculated from a sample of data. They are used to summarize and describe the characteristics of the sample. In essence, statistics are estimates or approximations of corresponding population parameters.

Sample-Based: Statistics are derived from a subset of data known as a sample, which is typically a smaller and more manageable subset of a larger population.

Variability: Since statistics are based on samples, they can vary from one sample to another, especially when dealing with small samples.

Purpose: Statistics are primarily used to analyze and draw conclusions about the sample from which they are collected. They provide insights into the sample's central tendencies, variability, and distribution.

Example of Statistics

Imagine you want to determine the average age of students in a school district with thousands of students. To make it more manageable, you randomly select 200 students from various schools and calculate the average age of this sample. This average age is a statistic, providing an estimate of the average age of students in the entire school district.

Statistic: The average age of the 200 randomly selected students is 15.5 years.

ii) Parameter

Definition: Parameters are numerical values that describe specific characteristics or attributes of an entire population. Unlike statistics, which are sample-based, parameters are population-based constants.

Population-Based: Parameters are ideally computed using data from an entire population. However, in practice, it may be challenging or impractical to collect data from an entire population; so parameters are often estimated using sample data.

Constant: Parameters do not vary; they remain fixed and consistent for the entire population under consideration.

Purpose: Parameters are used to describe and understand the population as a whole. They are often the ultimate goals of statistical analysis, as they provide a comprehensive view of the entire group of interest.

Example of Parameter

Suppose you want to know the precise average income of all households in a country with millions of households. Ideally, you would collect income data from every single household in the country to calculate the true parameter.

Parameter: The actual average income of all households in the country is Rs. 15,000.

1.9 Types of Variables

Variables in statistics can be classified into various categories based on different bases or characteristics. Here are some common classifications of variables based on different bases:

1. Based on the Measurement Scale

On the basis of scale of measurement, variables are of four types. These are as follows:

i) **Nominal Variables:** Nominal variables are categorical variables that represent distinct categories or groups with no inherent order or ranking. Examples include gender, ethnicity, skin colour, religion, marital status, etc. These variables can be used for classification or grouping purposes. Nominal variables with only two categories are called Dichotomous or Binary variables. On the contrary, Nominal variables with more than two categories are known as Polytomous variables.

ii) **Ordinal Variables:** Ordinal variables are also categorical variables but they have a meaningful order or ranking among their categories and the intervals between the categories are not uniform in nature. Examples include education level (e.g., high school, bachelor's degree, master's degree, and doctoral degree) or customer satisfaction ratings (e.g. very dissatisfied, dissatisfied, neutral, satisfied, and very satisfied).

iii) **Interval Variables:** Interval variables are numerical variables that have a consistent interval or difference between their values, but they lack a true zero point. Temperature is a classic example of an interval variable. The difference between 20°C and 30°C is the same as the difference between 30°C and 40°C. However, zero degrees Celsius (0°C) does not represent the complete absence of temperature. It is just an arbitrary point on the scale. Similarly, the pH scale measures the acidity or alkalinity of a solution. While the difference between pH 5 and pH 6 is meaningful and consistent with the difference between pH 6 and pH 7, a pH of 0 does not imply the complete absence of acidity.

iv) **Ratio Variables:** Ratio variables are numerical variables that have a consistent interval between their values and a meaningful zero point. This means that a value of zero indicates the complete absence of the attribute being measured. Examples include age, income, height, weight, number of items sold, etc. You can perform meaningful mathematical operations like addition, subtraction, multiplication, and division on ratio variables.

2. Based on Nature

On the basis of the nature of variables, variables are of two types. These are as follows:

i) **Categorical Variables:** Categorical variables, also known as qualitative variables, are a type of variable in statistics that represent data that can be divided into distinct categories or groups. These variables are used to categorize data into non-numeric groups or classes. Nominal and ordinal variables that represent categories or groups. Here are some examples of categorical variables. a) Gender is a categorical variable because it assigns individuals to distinct, non-numeric categories. In this case, the categories are typically 'male', and 'female'. b) Marital status is another categorical variable with categories such as 'single', 'married', 'divorced', and 'widowed'. These categories represent different marital statuses and do not have inherent numerical values. c) Education level is often categorized into groups such as 'high school', 'bachelor's degree', 'master's degree', and 'Ph.D.'. These categories represent different levels of education and do not have numeric values associated with them.

ii) **Numerical Variables:** Numeric variables, also known as quantitative variables, are a type of variable in statistics that represent data with numerical values that can be measured or counted. These variables can take on a wide range of numerical values, and they are used to represent quantities, measurements, or counts. Interval and ratio variables that represent quantities or measurements. Examples include age, income, height, weight, temperature, etc.

3. Based on Countability

On the basis of the countability, variables are of two types. These are as follows:

i) **Discrete Variables:** Discrete variables are variables that can only take on specific, distinct values within a finite or countable range. The number of children in a family or the number of cars in a parking lot are examples of discrete variables.

ii) **Continuous Variables:** Continuous variables are numerical variables that can theoretically take on an infinite number of values within a given range. Examples include height, weight, temperature, and income when measured with great precision. Continuous variables are typically measured, not counted.

4. Based on Demography

Demographic variables are characteristics or attributes that describe a population or a group of people. These variables are often used in research, marketing, and various other fields to analyze and understand different aspects of a population. Here are some common demographic variables along with examples:

i) **Age:** Example: 35 years old, 35-45 age group, senior citizen (65+).

ii) **Gender:** Example: Male, female, and transgender.

iii) **Race or Ethnicity:** Example: African, American, Asian, European, etc.

iv) **Education Level:** Example: High school, diploma, bachelor's degree, master's degree, Ph.D., etc.

v) **Income:** Example: $40,000 per year, low-income household, high-income household, etc.

vi) **Marital Status:** Example: Single, married, divorced, and widowed.

vii) **Household Size:** Example: 2-person household, family of four, living alone, etc.

viii) **Geographic Location:** Example: Urban, suburban, rural, specific city or region, etc.

ix) **Occupation:** Example: Teacher, software engineer, healthcare worker, etc.

x) **Religion:** Example: Christianity, Islam, Hinduism, Buddhism, etc.

xi) **Language:** Example: English, Spanish, Russian, French, Arabic, etc.

xii) **Family Structure:** Example: Nuclear family, extended family, single-parent household, etc.

xiii) **Disability Status:** Example: Physical disability, visual impairment, hearing impairment, etc.

xiv) **Immigration Status:** Example: Citizen, permanent resident, refugee, undocumented, etc.

xv) **Generation:** Example: Baby boomers, Generation X, Millennials, Generation Z, etc.

xvi) **Homeownership:** Example: Homeowner, renter, living with parents, etc.

xvii) **Sexual Orientation:** Example: Heterosexual, homosexual, bisexual, and asexual.

xviii) **Socioeconomic Status:** Example: Lower class, middle class, upper class, etc.

xix) **Health Status:** Example: Excellent, good, fair, poor, etc.

xx) **Family Size:** Example: Number of children in a family (e.g., one child, two children, etc.).

5. Based on the Nature of Data Collection

Here are explanations of each type of variable along with examples:

i) **Cross-Sectional Variables:** Cross-sectional data is collected from individuals, subjects, or entities at a single point in time. It provides a snapshot of a population at a specific moment. Cross-sectional variables describe the characteristics of different individuals or entities within a population at that specific moment.

Examples: a) Income of individuals surveyed in a random sample in 2022. b) The educational attainment of a group of students in a particular school district as of a specific date. c) Customer satisfaction scores collected from various customers during a single week.

ii) **Longitudinal Variables:** Longitudinal data is collected from the same individuals, subjects, or entities over an extended period, typically at multiple time points. It allows researchers to track changes and trends within the same group over time. Longitudinal variables describe changes or developments within the same individuals or entities across different time points.

Examples: a) Annual income of the same group of employees measured over a span of 10 years. b) Academic performance of students tracked from kindergarten through high school. c) Health data collected from patients at regular intervals over the course of a clinical trial.

iii) **Panel Variables:** Panel data combines both cross-sectional and longitudinal data. It involves collecting data from the same individuals, subjects, or entities at multiple time points. Panel variables include characteristics of the same individuals or entities over time, allowing for a more in-depth analysis of changes and trends within the same group.

Examples: a) A survey of households is conducted every quarter to track changes in income, expenses, and savings. b) Sales data of a particular product collected monthly from the same group of stores for two years. c) Employee performance evaluations conducted annually for the same group of workers over five years.

6. Based on Role in Analysis

Here are explanations of each type of variable along with examples:

i) **Independent Variable:** An independent variable is a variable in an experiment or research study that the researcher manipulates or controls to observe its effect on another variable, known as the dependent variable. The independent variable is the presumed cause, and researchers change or manipulate it to see how it influences the dependent variable.

Example: Studying the Impact of Study Time on Exam Scores

Independent Variable (IV): Study Time

This is the factor that the researcher is manipulating or varying. They might create different groups of students who are assigned different amounts of study time. For example, one group might study for 4 hours, another for 6 hours, and a third group for 8 hours.

Dependent Variable (DV): Exam Scores

This is the outcome or result that the researcher is measuring. In this case, it is the scores that the students receive on the exam. The researcher wants to see if the variation in study time (the independent variable) has an impact on these exam scores (the dependent variable).

ii) **Dependent Variable:** A dependent variable is a variable in an experiment or research study that is observed, measured, or analyzed to determine the effect of the independent variable. It is the outcome or result that researchers are interested in, and it depends on changes in the independent variable. In other words, the dependent variable is the variable that researchers seek to explain or predict based on the manipulation of the independent variable.

Example: Studying the Impact of Fertilizers on Plant Growth

Independent Variable (IV): Amount of Fertilizer

This is the factor that the botanist is manipulating or varying. They might create different groups of plants and provide them with different amounts of fertilizer. For example, one group might receive no fertilizer, another might receive a low amount of fertilizer, and a third group might receive a high amount of fertilizer.

Dependent Variable (DV): Plant Height

This is the outcome or result that the botanist is measuring. In this case, it is the height of the plants. The botanist wants to see if the variation in the amount of fertilizer (the independent variable) has an impact on the height of the plants (the dependent variable).

iii) **Mediating Variable:** A mediating variable, also known as an intermediate variable or mediator, is a variable that helps to explain the process or mechanism through which an independent variable affects a dependent variable. In other words, it provides insight into why or how the independent variable influences the dependent variable. Mediating variables are commonly used in research to better understand the underlying mechanisms at play in a relationship between two variables. Here is an explanation of mediating variables with an example.

Example: Mediating Variable in the Relationship between Stress and Physical Health

Let us consider a study that aims to understand how stress (independent variable) affects physical health (dependent variable) and how the perception of social support (mediating variable) plays a role in this relationship.

Independent Variable (IV): Stress

Stress is the factor that the researcher is interested in. It may be measured using a stress scale or questionnaire that assesses an individual's stress levels.

Dependent Variable (DV): Physical Health

Physical health represents the outcome of interest. It could be measured using various indicators, such as overall health ratings, the presence of physical symptoms, or medical assessments.

Mediating Variable (Mediator): Perception of Social Support

The mediating variable helps to explain the relationship between stress and physical health. In this case, the mediator is the individual's perception of social support, which can be assessed through surveys or questionnaires.

iii) **Moderating Variable:** A moderating variable, often referred to as an interaction effect or moderator, is a variable in a statistical model or research study that influences the strength or direction of the relationship between an independent variable and a dependent variable. Moderating variables shed light on when and for whom a relationship is more or less strong. They help to identify conditions under which the independent variable has a different effect on the dependent variable. Here is an explanation of moderating variables with an example.

Example: Moderating Variable in the Relationship between Exercise and Weight Loss

Let us consider a study examining the relationship between the frequency of exercise (independent variable) and weight loss (dependent variable) while considering age (moderating variable) as a potential moderator.

Independent Variable (IV): Frequency of Exercise

The independent variable in this study is the frequency of exercise, which can be measured as the number of days per week individuals engage in physical activity.

Dependent Variable (DV): Weight Loss

Weight loss represents the outcome of interest, and it can be measured in pounds or kilograms lost over a specified period.

Moderating Variable (Moderator): Age

Age is the moderating variable in this example. It is considered a moderating variable because researchers want to explore whether age influences the relationship between frequency of exercise and weight loss.

iv) **Control Variable:** A control variable, also known as a constant variable or covariate, is a variable in an experiment or research study that is intentionally kept constant or controlled to prevent it from affecting the relationship between the independent variable(s) and the dependent variable. Control variables are used to isolate and examine the specific effects of the independent variable(s) on the dependent variable while minimizing the influence of other potential factors. Here is an explanation of control variables with an example.

Example: Impact of Temperature on Plant Growth

Suppose a botanist wants to investigate how temperature (independent variable) affects the growth of a specific type of plant (dependent variable). In this experiment, several factors can influence plant growth, and the researcher wants to control some of them to focus solely on the impact of temperature.

Independent Variable (IV): Temperature

The independent variable is temperature, and the researcher wants to manipulate it to observe its effect on plant growth. Different groups of plants will be exposed to various temperature levels.

Dependent Variable (DV): Plant Growth

Plant growth is the outcome of interest, and it can be measured in terms of height, number of leaves, or other growth indicators.

Control Variable (Controlled Factor): Soil Quality

Soil quality is a control variable in this experiment. It is a factor that can affect plant growth but is held constant or controlled. All groups of plants are grown in identical soil conditions to ensure that soil quality does not vary and does not confound the results.

By controlling soil quality as a variable, the researcher aims to eliminate the possibility that differences in plant growth are due to variations in soil conditions rather than temperature.

v) **Confounding Variable:** A confounding variable, also known as a confounder or confounding factor, is an extraneous variable in a research study that is related to both the independent variable(s) and the dependent variable. It can lead to a wrong or misleading interpretation of the relationship between the independent and dependent variables because it introduces an alternative explanation for the observed effects. Identifying and controlling for confounding variables is essential to ensure the accuracy and validity of research findings. Here is an explanation of confounding variables with an example.

Example: Studying the Relationship between Coffee Consumption and Health of Heart

Suppose a researcher is interested in examining the relationship between coffee consumption (independent variable) and the risk of heart disease (dependent variable). The researcher collects data from a group of participants to investigate this relationship.

Independent Variable (IV): Coffee Consumption

The independent variable is the amount of coffee consumed by each participant. The researcher categorizes participants into three groups: low coffee consumption, moderate coffee consumption, and high coffee consumption.

Dependent Variable (DV): Risk of Heart Disease

The dependent variable is the risk of heart disease, measured as the possibility of developing heart-related health issues.

Confounding Variable (Confounder): Smoking Status

Smoking status is a confounding variable in this study. It is related to both coffee consumption and the risk of heart disease. People who consume more coffee are more likely to smoke, and smoking is also known to increase the risk of heart disease.

1.10 Role of Various Statistical Organizations

India has several statistical organizations that play crucial roles in collecting, analyzing, and disseminating statistical data. These organizations are helpful in providing reliable and up-to-date information for policy formulation, economic planning, research, and decision-making. These statistical organizations collectively play a crucial role in collecting, analyzing, and disseminating data that support evidence-based policymaking, economic planning, and research in India. They help to ensure that decisions made by governments, businesses, and other stakeholders are well-informed and grounded in reliable data.

Here is an overview of some of the key statistical organizations in India and their roles:

i) **Central Statistical Office (CSO):** The CSO, under the Ministry of Statistics and Programme Implementation, is the apex statistical agency in India. It is responsible for the compilation and publication of national accounts, including Gross Domestic Product (GDP) estimates. It also conducts the decennial Population Census and various economic and industrial surveys.

ii) **National Sample Survey Office (NSSO):** The NSSO, also under the Ministry of Statistics and Programme Implementation, conducts large-scale sample surveys to collect socio-economic data. It gathers data on various aspects of the Indian population, including employment, consumption, education, and health. The results of these surveys are used for policy formulation and program evaluation.

iii) Reserve Bank of India (RBI): The RBI collects and publishes data related to the financial and banking sector. It provides information on monetary policy, banking statistics, foreign exchange reserves, financial stability, etc. These data are essential for policymakers, economists, and financial institutions.

iv) National Crime Records Bureau (NCRB): The NCRB, under the Ministry of Home Affairs, compiles and publishes crime statistics for the country. It collects data on various types of crimes, including crimes against women, property crimes, and cybercrimes. These statistics aid law enforcement agencies and policymakers in understanding crime trends and taking appropriate actions.

v) Directorate of Economics and Statistics (DES) in States: Each state in India has its own Directorate of Economics and Statistics. These directorates are responsible for collecting and maintaining state-level statistical data, including data related to agriculture, industry, and demographics. They support State Governments in planning and development activities.

vi) National Informatics Centre (NIC): The NIC plays a critical role in developing and maintaining the technological infrastructure for data dissemination. It hosts and manages various government websites and portals that provide access to statistical data and information to the public, researchers, and policymakers.

vii) Indian Statistical Institute (ISI): ISI is an academic and research institution that conducts research in statistics and related fields. ISI contributes to the development of statistical methodologies and provides training and education in statistics and data science. It also collaborates with government agencies and organizations on research projects.

viii) National Institution for Transforming India (NITI Aayog): It is not exclusively a statistical organization. NITI Aayog plays a significant role in data-driven policy formulation. It uses statistical data and research to advise the government on various policy matters, including economic planning, social development, sustainable growth, etc.

1.11 Summary

Statistics is a branch of mathematics and a field of study that deals with collecting, organizing, analyzing, interpreting, and presenting data. The scope and functions of statistics are vast and multifaceted, encompassing various aspects of data collection, analysis, interpretation, and application. Statistics is a powerful tool for analyzing and interpreting data, but it also has certain limitations and constraints that should be recognized. Statistics is of two types. These are descriptive statistics and inferential statistics. Descriptive statistics involve summarizing and presenting data in a meaningful way. This includes measures of central tendency (e.g., mean, median, mode), measures of dispersion (e.g., standard deviation, range), and graphical representations (e.g., histograms, pie charts). Inferential statistics are used to make predictions, draw inferences, and test hypotheses about populations based on sample data. Techniques include hypothesis testing, confidence intervals, and regression analysis. Statistics and parameters are two different concepts. The main difference is that statistics are calculated from sample data and are used to estimate or describe population parameters, whereas parameters are fixed values that describe the entire population. Statistical organizations play a crucial role in collecting, analyzing, and disseminating statistical data and information. Their primary function is to provide accurate, reliable, and up-to-date data on various aspects of a country's economy, society, and environment.

1.12 Self-Assessment

1. What do you mean by statistics?
2. Discuss, in brief, the functions of statistics.
3. Mention the various scopes of statistics.
4. Highlight the limitations of statistics.
5. How the parameters are different from statistics?
6. What are the objectives of descriptive statistics?
7. Why is inferential statistics required?
8. Discuss various types of variables with suitable examples.
9. Explain the concept of the control variable with a suitable example.
10. Discuss the role of various statistical organizations in India.

Unit-II

Scale of Measurement and Data Structure

2.1 Learning Objectives

After completing this chapter, you should be able to:

i) Understand the concept and importance of nominal, ordinal, interval, and ratio scales;

ii) Identify the characteristics and uses of each scale type;

iii) Learn to select the right scale for different data and research needs;

iv) Recognize the properties and tests for each scale's adequacy;

v) Understand the challenges and limitations of using different scales;

vi) Stay updated on recent advancements in scale development;

vii) Apply nominal, ordinal, interval, and ratio scales in research with practical examples and calculations;

viii) Understand different types of data structure.

2.2 Introduction to Measurement in Statistics

In this chapter, we explore the fundamental principles of measurement scales in statistics. We start by understanding the concept and significance of nominal, ordinal, interval, and ratio scales. We then explore the characteristics and uses of each scale type, emphasizing the importance of selecting the appropriate scale for different types of data and research questions. Additionally, we discuss the properties and tests of adequacy for each scale, along with the challenges and limitations associated with their use. By staying informed of recent advances in scale development and measurement techniques, readers will be able to synthesize practical examples and calculations, enabling them to effectively apply these measurement scales in research and data analysis.

In the context of statistics, measurement refers to the process of giving variables or observations numerical values. The objective of this task is to quantify interesting traits, qualities, or characteristics. These numerical values allow the various aspects of the data to be methodically and consistently represented and contrasted. By doing this measurement, researchers and analysts are given the tools they need to interpret the data in meaningful ways, come to educated decisions, and carry out a range of statistical analyses. Not just physical measurements like length or weight are under the purview of statistical measurement. It broadens its scope to include arbitrary beliefs, habits, preferences, and other intangible phenomena that are amenable to measurement. Statisticians can use mathematical and analytical techniques to discover patterns, correlations, trends, and insights hidden in the data by assigning numerical values to these variables. The idea of measurement encompasses a variety of scales, each of which has unique properties and implications for analysis. These classes of measuring scales—nominal, ordinal, interval, and ratio—offer varied degrees of precision and usefulness to statistical studies. The inherent nature of the data and the specific goals guiding the analysis will determine the selection of an acceptable measuring scale.

2.3 Measurement's Function in Statistics

The importance of measurement is a fundamental concept in statistics that serves as the foundation for the field as a whole. In the same way that a map directs travellers through new territory, measurement offers the vital coordinates that lead us through the challenging terrain of data analysis. At its core, measurement is giving variables and observations numerical values in order to give data a quantitative quality. The ability to translate data into insights, identify patterns, and reach defensible conclusions depends on this transformation. Just as a symphony without a conductor would lack harmony and synchronisation among the different sounds, data remains dispersed and unstructured in the absence of measurement. Measurement arranges the data into a logical and standardised language that statistical algorithms can understand by giving numbers to the variables. With the use of this language, researchers, analysts, and decision-makers may have meaningful conversations with the data, uncovering its hidden meanings and transforming it into insights that can be put into practice.

In this investigation into the function of measurement in statistics, we explore the essential justifications for quantification's necessity. Measurement is the compass that directs us in our statistical endeavours, from revealing complex relationships to encouraging informed decision-making. As we explore measurement's importance, we learn about its implications for a variety of fields and how it enables us to navigate the challenging landscape of data analysis with assurance and accuracy.

2.4 The Primary Functions of Measurement in Research

The Primary Functions of Measurement in Research are listed below

a) **Quantification:** We may quantify and depict different aspects and characteristics of data using numerical values through measurement. This numerical form makes it possible to compare and analyse data points in a structured way.

b) **Standardisation:** By putting numbers next to each other, we establish a uniform framework for comprehending and evaluating data. It is now possible to share and transmit data-related information uniformly thanks to standardisation.

c) **Data Analysis:** Data analysis requires measurement as a precursor. Calculations, inferences, and meaningful conclusions from data are all achieved through the use of numerical values in statistical techniques and methodologies.

d) **Comparability:** It is simple to compare and contrast measured data. Statistical measurements help us to spot patterns, discrepancies, and similarities among data points so that we may make wise decisions.

e) **Finding Patterns and Interactions:** Data patterns, correlations, and interactions can be found more easily when using measurement. Understanding the underlying dynamics and trends evident in the data is crucial.

f) **Testing Hypotheses:** Statistical hypothesis testing compares observed data to anticipated results. The numerical foundation for conducting hypothesis tests and evaluating the significance of outcomes is provided by measurement.

g) **Making Informed Decisions:** Measurement enables us to glean insights from data, which supports decision-making. Data-driven judgements are more trustworthy and productive, whether they are made in business, research, or policy.

h) **Research and Science:** Scientific research relies on measurement to test hypotheses, support ideas, and advance knowledge. It would be difficult to establish empirical proof without measurement.

i) **Predictive Modelling:** In statistics, measurement serves as the foundation for the development of predictive models. These models use patterns in previous data to predict present and future trends, behaviours, and results.

j) **Quality Control:** Measurement is crucial for quality control in industries like manufacturing and product development. Organisations can ensure that items fulfil standards and requirements by measuring variables.

k) **Monitoring and Evaluation:** Measuring is essential for keeping track of progress and assessing results. It assists in gauging progress towards aims and goals and determining whether or not adjustments are required.

l) **Communication and Reporting:** Measurement gives data insights to many stakeholders in a common language. The communication of information is made simpler by using numerical quantities, whether in reports, presentations, or discussions.

Now let us look at some of the common areas where measurement is used.

2.5 Uses of Measurement

a) **Economic Indicators:** Measurement is essential for monitoring and evaluating the state of the economy in economics. The performance of economies is quantified using metrics like Gross Domestic Product (GDP), inflation rates, unemployment rates, and the Consumer Price Index. Governments, legislators, and corporations can develop plans and make educated decisions using these measurements.

b) **Medical Research:** To assess the efficacy of interventions and therapies, measuring is essential in medical research. To assess the effectiveness of drugs or treatments, clinical trials monitor things like blood pressure, heart rate, cholesterol levels, and patient-reported outcomes. These metrics inform medical standards and offer empirical support for technological improvements in medicine.

c) **Educational Assessment:** By evaluating student performance and learning outcomes, measurement plays a crucial role in educational research. The cognitive abilities of students are measured via standardised exams like the SAT or IQ testing. To analyse teaching methods and pinpoint areas for development, educational researchers also assess variables including student involvement, attendance rates, and grades.

d) **Market Research:** To understand consumer preferences and behaviour, market researchers use measurement. Numbers are gathered via surveys and questionnaires regarding consumer happiness, brand loyalty, and product preferences. These metrics aid companies in creating marketing plans, introducing new goods, and enhancing consumer experiences.

e) **Climate Science:** To follow environmental changes in climate research, measurement is crucial. To analyse climatic patterns and forecast future effects, variables including temperature, carbon dioxide concentrations, sea levels, and ice thickness are measured. These measurements help us comprehend the patterns of climate change and create measures to lessen its effects.

f) **Employee Engagement Assessment:** Measuring employee engagement is crucial in the fields of organisational behaviour and human resource management. Employee engagement, job happiness, and perceptions of the workplace are measured by surveys and questionnaires. These assessments give organisations information about the variables affecting engagement, allowing them to develop plans for enhancing workplace morale and productivity.

g) **Performance Appraisals:** Performance appraisal systems must include measurement. To evaluate employee performance, managers utilise quantifiable criteria like Key Performance Indicators (KPIs) and goals attained. Organisations can give useful feedback, pinpoint skill shortages, and offer specialised chances for training and growth by giving numerical numbers to different areas of performance.

h) **Diversity and Inclusion in the Workplace:** Measurement is essential for assessing these initiatives. Quantifying demographic information, including gender, ethnicity, and age, allows for analysing representation and equity in the workforce. Organisations can use these data to pinpoint discrepancies, monitor development, and carry out programmes that support diversity and inclusion.

i) **Talent Acquisition and Recruitment:** Measurement is crucial in human resource management during the talent acquisition process. Candidates' talents are measured through assessment exams and interviews, and resumes are evaluated for specific qualifications and experiences. Making educated hiring decisions, connecting candidates with job needs, and forecasting their prospective contributions to the organisation are all made easier with the use of these measurements.

k) **Training Effectiveness:** Measuring the effects of training and development initiatives is essential. Employee skill, knowledge, and behaviour improvements are measured before and after training. These metrics aid in evaluating the success of training programmes, allowing HR managers to improve training approaches and match them to organisational objectives.

The aforementioned describes how measurements are used throughout a range of disciplines and topics. The various examples show that measurement is the cornerstone upon which these fields construct plans, make judgements, and propel growth. Measurement equips professionals to analyse, contrast, and derive insights from the complex web of organisational dynamics and interpersonal interactions by turning abstract concepts into quantitative data.

2.6 Scales of Measurement

The diverse array of measurements encountered in different fields and contexts underscores the need for a variety of measurement scales. Each type of measurement, whether assessing employee engagement, economic indicators, or climate trends, requires a specific scale tailored to the nature of the data being quantified. Each type of measurement, whether assessing employee engagement, economic indicators, or climate trends, requires a specific scale tailored to the nature of the data being quantified. The wide-ranging types of measurements we encounter across various fields emphasize the significance of employing a diverse set of measurement scales.

Consider, for instance, the scenario of evaluating employee engagement. Here, we seek to gauge employees' emotional connection and commitment to their work. The scale used for this measurement might involve Likert-type questions, where respondents rate their agreement on a scale from "strongly disagree" to "strongly agree." This ordinal scale captures the ordinal relationship between responses, allowing us to discern the relative strength of engagement without implying equal intervals between each level of agreement.

Similarly, in the context of economic indicators, such as the unemployment rate or inflation rate, measurement scales play a crucial role. These variables require an interval scale, where the intervals between values are consistent and meaningful. For instance, if we have an inflation rate of 2% and another of 4%, we can ascertain that the latter has twice the inflationary impact, thanks to the consistent intervals on the scale. When investigating climate trends, measurement scales are equally vital. Suppose we're analysing temperature changes over time. Using a ratio scale, we can confidently assert that a temperature increase from 20° C to 40° C represents a doubling of heat energy, due to the presence of a true zero point on the scale. This enables us to accurately quantify and compare temperature changes.

In essence, the diversity of measurements necessitates a tailored approach to measurement scales. Each field demands a scale that aligns with the characteristics of the data under examination, ensuring accurate representation and meaningful interpretation. Just as a painter selects different brushes for various details, researchers and analysts choose the appropriate measurement scale to capture the nuances of the data they explore.

Scales play a critical role in data analysis as they determine the type of statistical analyses that can be applied to the data, influence the interpretation of results, and shape the conclusions drawn from the analysis. Different measurement scales impact data analysis in the following ways:

a) **Statistical Techniques:** The scale of measurement dictates which statistical techniques are appropriate for analysing the data. Each scale has its own set of permissible analyses. For instance, nominal scales might use chi-squared tests for association, ordinal scales might use non-parametric tests like the Mann-Whitney U test, interval scales allow for parametric tests like t-tests, and ratio scales enable more advanced analyses such as regression.

b) **Data Transformations:** Scales influence the need for data transformations. Variables measured on different scales may require transformations (e.g., logarithmic, square root) to ensure they meet the assumptions of certain analyses. Choosing the wrong transformation or failing to transform data appropriately can impact the validity of the results.

c) **Descriptive Statistics:** Scales determine the appropriate descriptive statistics. Nominal and ordinal scales might use mode and median, while interval and ratio scales can also incorporate mean and standard deviation. Using inappropriate descriptive statistics can lead to misleading interpretations.

d) **Comparisons:** Scales affect how data can be compared. Nominal scales allow for the identification of differences but not the magnitude of differences. Interval and ratio scales allow both comparison and quantification of differences.

e) **Interpretation:** The scale determines how results are interpreted. For instance, an increase of 10 units on an interval scale variable might imply the same increase in magnitude across the entire scale, while an increase of 10 units on a ratio scale implies a proportionate increase.

f) **Sampling and Hypothesis Testing:** The scale can influence sampling procedures and hypothesis testing. The appropriate sampling method might vary based on the scale, and the hypotheses formulated might differ based on the nature of the scale.

g) **Effect Size Calculations:** Different scales require different formulas to calculate effect sizes, which are important for quantifying the practical significance of findings.

h) **Graphical Representation:** The scale impacts how data is visualized. Nominal data might be presented in bar charts, ordinal data in box plots, and interval/ratio data in histograms or scatter plots.

i) **Level of Detail:** Scales determine the level of detail in the analysis. Nominal scales provide categorical distinctions, ordinal scales offer ordered distinctions, and interval/ratio scales provide both ordered and equal intervals.

j) **Model Assumptions:** Scales influence the assumptions that statistical models make. Violating these assumptions can affect the validity of the results.

2.7 Types of Scales

The various scale types namely, nominal, ordinal, interval, and ratio scales provide the necessary framework to categorize and analyse data with varying levels of precision. Just as a carpenter employs different tools for different tasks, the selection of an appropriate scale ensures that the nuances of each measurement are accurately captured, enabling effective interpretation and meaningful insights.

- Nominal Scale
- Ordinal Scale
- Interval Scale
- Ratio Scale

2.7.1 Nominal Scale

The nominal scale represents the foundational level of measurement wherein numerical values are employed as "tags" or "labels" to categorize and distinguish objects. This scale finds its relevance, particularly in cases involving non-numeric variables or situations where numbers lack inherent value. It entails the act of classifying variables based on qualitative descriptors, devoid of any inherent sequence or hierarchy, and without conveying any numerical value. To illustrate, consider the variable "hair colour," which could be segmented using a nominal scale into labels like blonde hair, brown hair, and grey hair. This segmentation does not indicate any quantifiable disparities or interrelationships among these labels. Further insights can be gained from an extensive exploration of nominal data.

In essence, a Nominal Scale operates as a tool of measurement wherein numbers function merely as markers to categorize or identify objects. This mode of measurement finds its primary utility in non-numeric variables or scenarios where numbers lack substantial meaning. An example of the Nominal Scale can be observed in the classification of data based on categories, with no inherent numerical hierarchy. Commonly referred to as both attribute and categorical data, nominal variables correspond to distinct groups defined by specific attributes. Although these groups can be labelled, they bear no intrinsic sequence or arrangement. In certain contexts, nominal data finds utility in grouping subsets of data, such as different experimental conditions within treatment and control groups, for the purpose of comparison.

a) **Gender Classification:** Gender is a classic example of a nominal scale. It involves categorizing individuals into distinct groups: male, female, and non-binary. These categories lack any inherent order or quantitative value; they simply serve as labels to classify individuals based on their gender identity.

b) **Blood Type Classification:** Blood types, such as A, B, AB, and O, are another example of a nominal scale. Each blood type represents a distinct category, and there is no meaningful order or hierarchy among them. Blood type classification is used in medical contexts for compatibility and transfusion purposes.

c) **Marital Status:** Marital status, including categories like single, married, divorced, widowed, and separated, exemplifies a nominal scale. These categories provide labels to classify individuals' relationship status, but they do not indicate any quantitative relationships or order between them.

d) **Eye Colour Categorization:** Categorizing eye colours, such as blue, brown, green, and hazel, is a nominal scale application. The categories represent different eye colour variations, but they lack a natural sequence or numerical value. This kind of categorization is often used in demographic studies and surveys.

e) **Vehicle Make and Model:** In the automotive industry, vehicle make and model can be classified using a nominal scale. Each make and model represents a distinct category (e.g., Toyota Camry, Ford Mustang). While these categories help identify different vehicles, they do not convey any intrinsic numerical or ordinal relationships.

The examples provided illustrate the application of nominal scales in various contexts, highlighting that the only information available is the distinct categories themselves. Whether classifying gender, blood type, marital status, eye colour, or vehicle makes and models, the focus remains solely on labelling and categorizing data points. These categories lack any inherent order or quantitative value, serving as fundamental identifiers without implying any numerical relationships. Nominal scales offer a structured framework for organizing data based on qualitative attributes, allowing for clear differentiation among categories without involving measurement in the traditional numeric sense.

2.7.1.1 Characteristics of Nominal Scale

a) **Categories:** Nominal scales consist of distinct categories or labels that represent different groups or classes. These categories are mutually exclusive and collectively exhaustive, meaning each observation falls into only one category, and all possible observations are covered by the categories.

b) **No Order:** Unlike other measurement scales, nominal scales do not possess any inherent order or sequence among the categories. The categories are simply names or labels and do not convey any information about the magnitude of differences between them.

c) **Mutually Exclusive:** Each observation can belong to only one category. There is no overlap or ambiguity in assigning observations to the respective categories.

d) **No Magnitude:** Nominal scales do not imply any quantitative or numerical relationships between categories. The differences between categories are not measurable or meaningful.

e) **No Arithmetic Operations:** Since nominal scales lack numerical values, arithmetic operations like addition, subtraction, multiplication, or division are not meaningful or applicable to the categories. Therefore, calculations involving nominal scale data are limited to counts and percentages.

2.7.1.2 Operations of Nominal Scale

a) **Categorization:** The primary operation of a nominal scale is categorizing observations into distinct groups based on their attributes. This categorization facilitates organizing and structuring data for analysis.

b) **Counting and Frequencies:** Nominal scale data can be counted to determine the frequency of observations in each category. This provides insight into the distribution of data across different groups.

c) **Mode Calculation:** The mode, which represents the most frequently occurring category, is a meaningful measure of central tendency for nominal scale data.

d) **Cross-tabulation:** Nominal data can be cross-tabulated to explore relationships between two or more nominal variables. This involves creating a contingency table that shows the joint frequency distribution of different categories.

e) **Chi-Squared Test:** Statistical tests like the chi-squared test can be applied to analyse associations between different nominal variables.

This test assesses whether observed frequency distributions differ significantly from expected distributions.

f) **Bar Charts and Pie Charts:** Nominal scale data is often presented using bar charts or pie charts, where the frequency or proportion of each category is visually represented.

g) **Comparative Analysis:** Nominal data can be used to compare the distribution of categories among different groups or subgroups. This can provide insights into disparities or patterns.

h) **Grouping and Summarizing:** Nominal data can be grouped into broader categories to simplify analysis or presentation. This is particularly useful when dealing with a large number of distinct categories.

2.7.2 Ordinal Scale

The ordinal scale represents a progression beyond the nominal scale, encompassing additional attributes. When the aim involves not just categorizing variables but also establishing a hierarchical rank among them, the ordinal scale comes into play. This scale introduces the concept of items possessing varying levels of hierarchy, wherein one item holds a higher position compared to another. For instance, consider the scenario where you solicit your friend's or survey respondents' rankings for five different types of toothpaste, assigning a rank of one to the most-used and a rank of five to the least-used toothpaste. This exemplifies the application of an ordinal scale. Unlike the nominal scale, ordinal scales incorporate a degree of order among the categories, allowing these categories to be naturally arranged in a sequence. While the order among categories holds significance, the extent of difference between them might not be uniformly consistent. Ordinal scales enable the ranking of values, yet the specific degrees of difference between these ranked values remain unspecified. Let us consider the following examples:

1. **Education Level:** The ordinal scale is applicable when categorizing education levels, such as "high school diploma," "bachelor's degree," "master's degree," and "doctorate degree." While these categories exhibit an ordered progression, the exact intervals between them are not uniform.
2. **Customer Satisfaction Ratings:** Assigning customer satisfaction ratings like "very dissatisfied," "dissatisfied," "neutral," "satisfied," and "very satisfied" on surveys employs an ordinal scale. The categories reflect varying levels of satisfaction, but the intervals between them might not signify consistent differences.

3. **Likert Scale Responses:** The Likert scale, often used in surveys, employs ordinal scaling. Respondents are asked to indicate their agreement levels using categories like "strongly disagree," "disagree," "neutral," "agree," and "strongly agree." A Likert scale is frequently employed to gauge attitudes, knowledge, perceptions, values, and behavioural shifts. It typically consists of a range of statements that respondents can select from to rate their responses to evaluative questions.
4. **Performance Ratings:** Employee performance evaluations often use ordinal scales, with categories like "below expectations," "meets expectations," "exceeds expectations," and "outstanding." These categories reflect varying levels of performance, but the exact differences between them might not be consistent.
5. **Star Ratings for Reviews:** Online reviews often employ ordinal scales in the form of star ratings, such as one to five stars. While a higher number of stars generally indicates a more positive review, the intervals between the star ratings do not necessarily signify equal levels of difference.

2.7.2.1 Characteristics of Ordinal Scale

1. **Order:** Ordinal scales have categories that possess a meaningful order or hierarchy. The categories represent different levels of a characteristic, indicating relative positions without specifying the magnitude of differences between them.
2. **No Equal Intervals:** While ordinal scales show order, the intervals between categories are not necessarily equal or consistent. The differences between categories are not standardized or quantifiable.
3. **Mutually Exclusive:** Similar to nominal scales, observations are assigned to a single category, and there is no overlap or ambiguity in categorization.
4. **No True Zero:** Ordinal scales lack a true zero point, which means there is no absolute starting point or origin. Ratios and arithmetic operations involving ordinal data are not meaningful.
5. **Limited Arithmetic Operations:** While some arithmetic operations like comparisons (greater than, less than) are possible, operations involving addition, subtraction, multiplication, or division are not suitable due to the absence of equal intervals.

2.7.2.2 Operations of Ordinal Scale

1. **Ranking:** The fundamental operation of an ordinal scale is ranking observations based on their position within the hierarchy of categories. This ranking provides insight into the relative order of observations.

2. **Median Calculation:** The median, representing the middle value in a ranked list, is a meaningful measure of central tendency for ordinal data. It considers the order of categories but not their magnitudes.
3. **Percentiles:** Percentiles can be calculated to understand the distribution of observations relative to the ordered categories.
4. **Comparisons:** Ordinal data allows for comparisons such as "greater than," "less than," or "equal to." However, the extent of the difference between categories is not quantifiable.
5. **Non-Parametric Tests:** Non-parametric tests like the Wilcoxon rank-sum test or the Kruskal-Wallis test are suitable for analysing differences between groups with ordinal data. These tests consider the ranked order of observations.
6. **Bar Charts and Dot Plots:** Graphical representations like bar charts and dot plots can be used to display the distribution of ordinal data and highlight the order among categories.
7. **Qualitative Descriptions:** Ordinal data can be summarized using qualitative descriptions of categories, such as "high," "medium," and "low," based on their position within the ordered scale.
8. **Cumulative Frequency Analysis:** Ordinal data can be analysed using cumulative frequency distributions, which show the number or proportion of observations falling below a certain category.

2.7.3 Interval Scale

The third tier of measurement is the interval scale, distinct for its numerical essence, offering capabilities surpassing those of nominal and ordinal scales. Within this scale lies the property of both order and distance. Unlike nominal and ordinal scales, the interval scale boasts a meaningful mathematical gap between values. However, it's crucial to note that this scale lacks a "true zero." In other words, zero on this scale doesn't imply a complete absence of the measured attribute. For instance, when the temperature reads '0' degrees, it doesn't signify the complete absence of temperature; it's simply a point of reference. Mathematical operations like addition, subtraction, mean calculation, and standard deviation computation can be performed on this scale, facilitating more in-depth analysis. Notable examples that fall within the realm of an interval scale include temperature, time, and date – all of which can be treated as interval data. Skilfully crafted, this scale can serve to measure perception and opinion, as exemplified by the diverse array of survey questions on platforms like SurveyMonkey. Furthermore, statistical tests can be applied to data derived from this scale, enhancing its analytical potential.

An interval scale marks a level of measurement that extends beyond nominal and ordinal scales, introducing a numeric dimension to the characterization of data. This scale possesses two distinctive properties: order and distance. In contrast to nominal and ordinal scales, which encompass categorical or ranked data, the interval scale incorporates a meaningful mathematical difference between values. However, it is important to recognize that an interval scale lacks a "true zero." Let us explore the following examples of interval scales.

a) **Temperature Measurement:** The measurement of temperature using the Celsius or Fahrenheit scale is a classic example of an interval scale. The intervals between temperature values are consistent and meaningful, enabling calculations like finding temperature differences or averages.

b) **IQ Test Scores:** IQ (intelligence quotient) test scores, which quantify cognitive abilities, are often treated as interval scale data. The consistent intervals between different scores allow for precise comparisons and calculations.

c) **Calendar Dates:** Calendar dates, represented by days, months, and years, are another example of an interval scale. The intervals between these units are constant and interpretable, facilitating calculations involving periods or durations.

d) **Economic Indices:** Economic indicators like inflation rates, GDP growth rates, and stock market indices are measured on interval scales. The differences between these numeric values hold significant economic implications, allowing for comparisons and analysis.

In these examples, the interval scale showcases both order and consistent intervals between values. This characteristic enables various mathematical operations, precise comparisons, and quantitative analysis, making interval scales essential in fields ranging from science and economics to social research and everyday life.

2.7.3.1 Characteristics of Interval Scale

a) **Order and Distance:** The interval scale features both order and distance between its values. Each value represents a specific level, and the intervals between values hold a consistent and interpretable numerical difference.

b) **Absence of True Zero:** Notably, the interval scale does not have a "true zero." In practical terms, this means that a zero value does not signify the complete absence of the attribute being measured. For instance, a temperature reading of 0°C does not indicate a lack of temperature; it represents a reference point.

2.7.3.2 Operations and Potential of Interval Scale

a) **Mathematical Operations:** An exceptional feature of the interval scale is its compatibility with mathematical operations like addition, subtraction, multiplication, and division. These operations are meaningful due to the consistent intervals between values.

b) **Statistical Measures:** Interval scales allow the calculation of essential statistical measures such as the mean and standard deviation. These measures provide valuable insights into the central tendency and dispersion of the data.

c) **Comparative Analysis:** Interval scales facilitate accurate comparisons, as differences between values hold quantitative significance. "Greater than," "less than," and "equal to" relationships are both interpretable and meaningful.

d) **Graphical Representation:** Interval scale data lends itself to various graphical representations, including histograms, scatter plots, and line charts. These visualizations enhance data interpretation.

e) **Precision in Measurement:** The consistent intervals between values contribute to the precision of measurements, enabling researchers to discern and communicate nuanced differences.

2.7.4 Ratio Scale

The ratio scale mirrors the interval scale closely, yet it distinguishes itself with a pivotal attribute: the presence of a "true zero." The pivotal distinction lies in the fact that on a ratio scale, a true zero point signifies an absolute absence of the attribute being measured. Consider the exemplar of weight in kilograms; if an object registers a weight of zero kilograms, it unequivocally implies an absence of weight. This differs from the realm of temperature (interval data), where a reading of zero degrees doesn't indicate an absence of temperature but rather extreme coldness. Some examples of ratio scales include:

a) **Weight in Kilograms:** The measurement of weight in kilograms exemplifies a ratio scale. A weight of zero kilograms signifies a complete absence of weight, and ratios between weight measurements are meaningful.

b) **Height:** Height measurements, such as in meters or feet, fall under the domain of a ratio scale. Ratios between different height measurements are meaningful and accurate.

c) **Time Periods:** Measurements involving time periods, like durations or intervals, adhere to the ratio scale. Ratios between time intervals, such as comparing a 10-minute duration to a 5-minute duration, are valid and interpretable.

d) **Speed:** Speed measurements, whether in kilometres per hour or metres per second, are represented on a ratio scale. Ratios between different speed values hold significant meaning.

e) **Temperature in Kelvin:** Temperature measurements in Kelvin, which feature an absolute zero that represents an absence of temperature, are categorized within the realm of a ratio scale.

2.7.4.1 Characteristics of Ratio Scale

a) **True Zero:** The hallmark of the ratio scale is the existence of a true zero, setting it apart from the interval scale. On a ratio scale, a value of zero holds the significance of the absolute nonexistence of the measured attribute.

b) **Continuous Data:** Ratio scales are often associated with continuous data, where intervals remain meaningful and measurements can adopt any real value within a given range.

c) **Validity of Ratios:** This scale type enables not only the ordering of observations and assessment of differences but also the examination of ratios between values. For instance, 30 kg. being thrice the weight of 10 kg. underscores the scale's ability to represent and interpret ratios accurately.

The ratio scale is a powerful measurement framework that enables a range of operations and analytical possibilities due to its unique properties. Here's a deeper exploration of its operations and potential:

2.7.4.2 Operations and Potential of Ratio Scale

a) **Arithmetic Operations:** The ratio scale's order, consistent intervals, and true zero enable seamless arithmetic operations like addition, subtraction, multiplication, and division. This is crucial for quantitative analysis and scientific investigations.

b) **Ratios and Proportions:** The ratio scale excels in assessing ratios and proportions, offering insights into relative quantities. Fields like engineering, physics, and finance benefit from this feature.

c) **Statistical Measures:** Similar to interval scales, the ratio scale supports statistical measures such as mean, median, mode, standard deviation, and variance, aiding in understanding data distribution.

d) **Meaningful Ratios:** Calculating and interpreting ratios between values enhances analysis depth, enabling precise comparisons and insights into relationships.

e) **Informed Decision-Making:** The ratio scale's suitability for quantitative analysis is vital in making accurate decisions, benefiting industries like manufacturing and research.

f) **Scientific Modelling:** The ratio scale's ability for arithmetic operations and ratio assessment enriches scientific modelling accuracy, aiding in predictions and equations.

g) **Proportional Scaling:** The true zero allows proportional scaling of measurements, essential for converting data between units accurately.

h) **Precise Measurement:** The ratio scale's order, intervals, and true zero contribute to measurement precision, making it indispensable for experiments, research, and data-driven decisions.

Table 2.1: Illustration of Data Types and Scale Types with Examples

S No.	Examples of Data	Answer
1	Gender (Male, Female)	Categorical- Nominal scale
2	Customer satisfaction rating (1-5)	Discrete- Ordinal
3	Temperature in Celsius	Continuous-Interval scale
4	Income in dollars	Continuous-Ratio scale
5	Educational qualification (High School, Bachelor's, Master's, PhD)	Discrete-Ordinal;
6	Number of employees in a company	Discrete-Interval scale
7	Stock prices in a financial market	Continuous-Ratio scale
8	Age of participants in a study	Continuous- Ratio scale
9	Customer loyalty level (Low, Medium, High)	Discrete -Ordinal
10	Time taken to complete a task	Continuous; Interval scale (Zero represents a starting point, not the absence of the attribute)

As a diagrammatic summary level of measurement can be can be viewed as:

The Four Levels of Measurement:

	Nominal	Ordinal	Interval	Ratio
Catagorizes and labels variables	✓	✓	✓	✓
Ranks categories in order		✓	✓	✓
Has known, equal intervals			✓	✓
Has a true or meaningful zero				✓

Fig 2.1: Permissible Operations in Various Scales

2.8 Significance of Different Scales

The diverse array of measurement scales - nominal, ordinal, interval, and ratio - holds paramount importance in the process of quantifying and comprehending data across a multitude of domains. Each scale brings its distinct attributes, serving as a vital tool for capturing the inherent characteristics of data. Their significance is best highlighted through their practical implications in real-world applications. Nominal scales establish a fundamental groundwork for the categorization and identification of qualitative attributes, enabling data organization and segmentation in areas like market research. Ordinal scales introduce order and hierarchy among categories, allowing for meaningful comparisons and rankings. This proves essential in domains ranging from education, where student performance is ranked, to evaluating consumer preferences and satisfaction. Interval scales, characterized by consistent intervals between values, permit arithmetic operations and precise analysis. These scales find practical utility in fields as diverse as engineering, where precise temperature control is vital, and in survey research, where they offer insights into the strength of attitudes. Ratio scales, encompassing a true zero, excel by enabling meaningful ratios and proportions. They play a pivotal role in medical diagnosis, financial planning, and scientific research, where accurate measurement is paramount. In practice, these scales contribute to data accuracy, facilitate comparative analysis, and determine the appropriate statistical tests for validity. They enable personalization, efficient communication of trends, and informed decision-making across an array of sectors. Collectively, the significance of these measurement scales lies in their capacity to transform raw data into meaningful insights, ultimately shaping our understanding of diverse phenomena and enhancing our ability to navigate complex information landscapes.

In essence, summarizing the key takeaways, when the aim of measurement revolves around categorization, the appropriate choice is the nominal scale. For instance, distinguishing between male and female falls under this category. In scenarios where the intention is to establish a ranking among variables or different attributes, an ordinal is employed. Should the objective shift to evaluating the differences between two intervals of continuous nature, or necessitate the utilization of statistical measures like mean and standard deviation for data comprehension, the interval scale becomes the suitable option. However, it is important to note that this scale lacks a true zero point. On the other hand, when the purpose involves measuring continuous variables while retaining the crucial attribute of a true zero on the scale, the ratio scale comes into play. This scale proves its utility in scenarios where the presence of a true zero holds significance, allowing for precise quantification and comparison of data.

2.9 Some Important Scales

a) **Ranking Scale:** A measurement technique used in research to determine the relative importance or preference of several things or possibilities is the ranking scale. With this scale, participants must contrast and group items according to how important, valuable, or relevant they are in their opinion. Typically, respondents are shown pairs of objects and asked to rank one item higher than the other to indicate their preference. This procedure is repeated for numerous combinations, producing a ranked list that eventually reflects the preferences or priorities of the participants. In domains like product preference studies, marketing research, and decision-making analyses, ranking scales are particularly helpful because they help us comprehend the relative relevance of objects within a specific context.

b) **Rating Scale:** It is a sort of measurement tool used in surveys and research to quantify respondents' subjective traits, beliefs, attitudes, and perceptions. Rating scales are useful instruments for analysing opinions, preferences, and attitudes in a variety of domains, including psychology, market research, education, and social sciences. They allow researchers to extract quantitative data from qualitative notions. The selection of a rating scale is based on the goals of the study, the kind of information being gathered, and the environment in which the assessment is being done. Participants can express their preferences, opinions, or evaluations within a structured framework using a scale that corresponds to different levels of the measured property. The degree of agreement, satisfaction, importance, or intensity connected to certain statements, concepts, products, or experiences is captured through rating systems. There are various categories of rating scales, each with a specific function:

c) **Categorical Rating Scale:** The simplest rating scale is the dichotomous one, which just offers two alternatives for responses, such as "Yes" or "No," "True" or "False." Participants choose from predetermined categories, such as gender ("Male," "Female") or academic year ("First year," "Second year," etc.).

d) **Semantic Differential Scale:** This scale uses pairs of opposing adjectives to rate an item, concept, or idea. It captures the evaluative dimension between the pairs of adjectives (e.g., "Good" to "Bad," "Friendly" to "Unfriendly"). This scale asks respondents to rate an object, concept, or event on a set of bipolar adjectives that reflect different dimensions. For example, participants might rate a product on attributes like "expensive" to "affordable" or "unreliable" to "reliable.

e) **Itemised Rating Scale:** Respondents rate a number of characteristics or assertions by choosing one of several possible responses for each item.

f) **The Stapel Scale:** It allows participants to score items or concepts along a bipolar continuum by placing a single adjective, often in the scale's middle. A Stapel scale is an adjective-only, closed-ended rating system that is generally created to elicit responses from the respondents of a study regarding a certain topic or event. There are an even number of possible answers to the survey question, but there is no neutral answer. These possibilities, for instance, lack a neutral point and can vary from any scale, such as +2 to -2, +5 to -5, or even +10 to -10, depending only on the nature of the study and the desired outcome. The Stapel scale can be used as solo questions or even as a matrix question type with one distinct adjective listed for each line item.

+5	+5
+4 X	+4
+3	+3 X
+2	+2
+1	+1
High Food Quality	Poor Crew Service
-1	-1
-2	-2
-3	-3
-4	-4
-5	-5

Fig 2.2 Example of a Stapel Scale

g) **Graphic Rating Scale:** In order to express their level of agreement, satisfaction, or other factors, participants place a mark on a line or scale, creating a visual representation of their response. Participants use the consensus scale to indicate how much they agree or disagree with a statement based on the apparent consensus of a group or team.

h) **Likert Scale:** A Likert scale is a common type of rating scale used to measure attitudes or opinions. Respondents are asked to indicate their level of agreement or disagreement with a statement using a range of response options (e.g., "strongly agree," "agree," "neutral," "disagree," "strongly disagree"). The Likert scale is a commonly employed evaluative framework in academic research and survey methodologies, designed to assess subjective attitudes, views, or perceptions. The survey instrument provides participants with a set of statements accompanied by a variety of response alternatives that reflect different degrees of agreement or disagreement. These options often include choices such as "Strongly Disagree," "Disagree," "Neutral," "Agree," and "Strongly Agree." The responses are allocated numerical values, facilitating quantitative analysis. The Likert scale is a helpful tool due to its simplicity, ability to be quantified, adaptation to different topics, and comparability between respondents. The need for precise language and thoughtful design of answer options cannot be overstated in order to provide meaningful and dependable data collection. It is also essential to recognise potential limitations such as response bias and the assumption of equal intervals between alternatives.

These scales play a crucial role in research, as they provide a standardized way to measure and analyse variables, enabling researchers to draw meaningful conclusions from their data. The choice of scale depends on the nature of the variable being measured and the research objectives.

2.10 Data Structure

A data structure is a specialised framework that is employed to efficiently organise, modify, store, and retrieve data. A wide range of data structure types, encompassing both fundamental and advanced concepts, have been devised to serve distinct objectives in the organisation of data. These structures enable efficient user access and interaction with necessary data, assuring the suitability of their utilisation. Data structures play a crucial role in providing a framework that enables effective understanding of information by both people and robots. The design of data structures is intricately intertwined. Every data structure contains essential information. The importance of data structures arises from the limitations of fundamental data types, such as integers or floating-point values, in effectively representing the complex logical aspects of data manipulation and utilisation. In the context of application functionality involving the ingestion, manipulation, and generation of information, it is crucial to comprehend the most effective approach to organising data in order to enhance processing efficiency. Data structures are entities that combine data elements in a cohesive manner, hence enabling efficient utilisation, durability, and sharing of data. The authors present a structured framework that delineates the arrangement of data components. Data structures are essential aspects of complex systems. They are created by combining data pieces into logical units that represent abstract data types relevant to certain algorithms or applications. One possible example of an abstract data type may be demonstrated through the amalgamation of character strings representing "first name," "middle name," and "last name," so constituting the abstract notion of a "customer name." The process of choosing data structures extends beyond simple usage and requires the careful selection of the most suitable data structure for a specific activity. Selecting an inappropriate option may result in decreased programme execution speed or unresponsiveness of the code. When making a decision about which data structure to use, several important factors should be taken into consideration.

i) The nature of the information to be stored.
ii) How the information will be employed.
iii) Determining the suitable location for data persistence after its creation.
iv) Identifying the optimal manner to organize the data.
v) Contemplating memory and storage reservation management aspects.

Thus, data structure assumes a fundamental role, representing the architecture through which data is organized, stored, and presented within a dataset. The selection of an appropriate data structure holds pivotal significance, as it profoundly influences the efficiency and effectiveness of data management, analysis, and interpretation. Diverse data structures boast distinct attributes that exert a significant impact on how data is accessed, manipulated, and processed. From arrays that enable the arrangement of elements in a linear sequence, facilitating swift and direct access through indices, to lists that accommodate varying data types and sizes dynamically, granting flexibility in insertion and deletion operations, researchers meticulously opt for data structures aligned with their specific research contexts. Linked lists, characterized by their utilization of nodes that encapsulate data and references to subsequent nodes, exhibit prowess in dynamic data storage scenarios where frequent changes are prevalent. The utility of stacks and queues, guided by the principles of Last-In-First-Out (LIFO) and First-In-First-Out (FIFO) respectively, extends to tasks as diverse as function call management, undo/redo operations, and algorithmic traversals. Trees, imbued with hierarchical node relationships, are adept at representing various relationships and hierarchies within data, while graphs, composed of interconnected vertices, elucidate intricate relationships among entities. Hash tables expedite data retrieval by employing hash functions to map keys to indices, and matrices find their niche in statistical analysis and linear algebra. Databases, the bedrock of structured data storage, deploy a plethora of data structures to ensure the efficient storage and retrieval of information. The meticulous selection of a data structure hinges on myriad factors, including the nature of the data, the research objectives, computational resources, and the operations required. In essence, the choice of a data structure is a strategic decision that significantly shapes the course of data analysis, interpretation, and the overarching success of research endeavours.

2.10.1 Cross-Sectional Data

Cross-sectional data refers to information collected from diverse subjects or entities at a specific point in time, offering a snapshot view of characteristics, behaviours, or variables within a given population. This type of data enables researchers to analyse relationships, differences, and patterns across various entities simultaneously, making it useful for comparative studies and exploring correlations. Cross-sectional data does not track changes over time and is particularly valuable when assessing the current status of a phenomenon or conducting quick and cost-effective analyses, although it may not provide insights into causal relationships or temporal developments.

Examples of cross-sectional data use-

i) Gathering cross-sectional data on educational attainment and income levels across neighbourhoods to explore their correlation.

ii) Analysing cross-sectional data from various industries to assess the impact of technology adoption on job roles.

iii) Using cross-sectional data to investigate the relationship between air quality indices and respiratory health in a city's population.

iv) Examining cross-sectional data of consumer preferences to understand how demographic factors influence purchasing behaviour.

v) Surveying cross-sectional data on political affiliations within a community to discern patterns of voter sentiment.

vi) Employing cross-sectional data to examine the prevalence of mental health conditions in different age groups.

Table 2.2 Example of a Cross Sectional Data

Participant ID	Age	Gender	Education Level	Income (INR)	Satisfaction Score
1	35	Male	Bachelor's	₹ 3,50,000	7
2	45	Female	Master's	₹ 5,60,000	8
3	28	Female	High School	₹ 2,45,000	5
4	55	Male	PhD	₹ 8,40,000	9
5	40	Male	Post Doc	₹ 4,20,000	6

In this table, each row represents a different participant, and each column represents a different variable (age, gender, education level, income, and satisfaction score). This type of data is collected at a single point in time and provides a snapshot of the participants' characteristics and responses.

2.10.1.1 Pros of Cross-Sectional Data

Cross-sectional data offers several advantages in research. Firstly, it is an efficient and cost-effective approach, involving data collection at a single point in time, which is particularly valuable when time and resources are limited. This method also enables researchers to gain diverse insights by simultaneously studying various variables, characteristics, or groups within a population. Cross-sectional data facilitates comparative analyses, allowing researchers to discern patterns, differences, and correlations across different entities. Moreover, this type of data provides a snapshot view of the current state of a phenomenon, aiding in describing and understanding the prevailing circumstances. Additionally, cross-sectional data serves as a crucial baseline for future studies, enabling researchers to track changes, trends, and developments over time by comparing subsequent data to the initial snapshot.

2.10.1.2 Cons of Cross-Sectional Data

Despite its advantages, cross-sectional data has limitations that researchers must consider. One significant drawback is the lack of temporal information; since data is collected at a single time point, it does not capture changes or trends over time, limiting insights into causal relationships and temporal dynamics. Inferences about causality can be challenging to establish solely based on cross-sectional data, as it does not provide conclusive evidence of cause-and-effect relationships. Additionally, the absence of temporal context might result in the omission of underlying factors that contribute to observed patterns. There is also a risk of selection bias, where the collected data coincides with specific events or conditions, potentially skewing the results and leading to an inaccurate representation of the usual situation. Consequently, while cross-sectional data offers valuable insights, researchers need to approach its interpretation cautiously and consider its limitations when drawing conclusions about causality or long-term trends.

2.10.2 Time Series Data

Time series data is a type of data where observations are recorded at specific time intervals. Each observation is associated with a timestamp, allowing for the analysis of how a particular variable changes over time. This data is often used in various fields such as economics, finance, environmental science, and more, to understand patterns, trends, and relationships that develop over time. Time series analysis involves techniques like trend analysis, seasonal decomposition, and forecasting to gain insights from this sequential data.

In time series data, observations are noted at predetermined intervals of time. Because each observation has a timestamp, it is possible to examine how a given variable changes over time. To analyse patterns, trends, and relationships that emerge over time, this data is frequently employed in a variety of disciplines, including economics, finance, environmental research, and more.

When working with data that has a time component, such as stock prices, weather measurements, sales data, and more, this form of analysis is quite helpful. In order to make predictions, forecasts, and decisions, time series analysis aims to find patterns, trends, and underlying structures in the data.

Here are some essential ideas and procedures for time series analysis:

1. **Time Series Components:** Time series data frequently includes a number of these elements:
   - **Trend:** The data's long-term motion or direction over time.
   - **Seasonality:** Repeated patterns or recurring swings across time.
   - **Cyclic Patterns:** Longer-term, less predictable oscillations than seasonal patterns.
   - **Unpredictable** fluctuations that don't adhere to a regular pattern are known as irregular or random variations.
2. **Data Pre-Processing:** It's important to clean and prepare the data. This includes dealing with missing numbers, smearing noisy data, and, if necessary, altering the data.
3. **Exploratory Data Analysis (EDA):** EDA entails visualising the data to comprehend its features and spot trends, seasonality, and any outliers or anomalies that can influence the analysis.
4. **Decomposition:** A time series is decomposed into its trend, seasonality, and residual (irregular fluctuations) components. This makes it easier to separate and examine each component separately.

5. **Smoothing Techniques:** Smoothing techniques are used to minimise noise and highlight underlying patterns in the data. Examples include moving averages and exponential smoothing.
6. **Stationarity:** Many time series analysis methods make the assumption that the data is stationary, i.e., that statistical constants like mean and variance do not change over the course of time. It could be necessary to alter non-stationary data to make it stationary.
7. **Lag Analysis and Autocorrelation:** Lag analysis examines the relationship between a data point and earlier data points at various time lags. It facilitates the discovery of recurring patterns.
8. **Modelling:** There are two basic categories of time series models:
   a) **ARIMA (Auto Regressive Integrated Moving Average):** This well-liked model combines moving average (MA) and autoregressive (AR) components, making it appropriate for stationary data.
   b) **Exponential Smoothing Models:** These models give more weight to recent data and different weights to historical observations.
9. **Forecasting:** A model can be used to predict future values once it has been constructed and validated. Metrics like Mean Absolute Error (MAE) or Root Mean Squared Error (RMSE) can be used to judge the model's correctness.
10. **Model Evaluation and Improvement:** Models should be assessed using out-of-sample data, and they should be compared to one another in terms of performance. Parameter adjustment and model testing may be involved in iterative improvement.

Time series analysis is a flexible tool with uses in many fields, including finance, economics, marketing, healthcare, and more. By comprehending and forecasting future trends based on historical data, it enables businesses and researchers to make informed judgements.

Table 2.3 Illustration of a Time-Series Data

Year	Table 2.3: Illustration of a Time-Series Data	Inflation Rate (%)	Unemployment Rate (%)	Interest Rate (%)	Stock Market Index
2016	5	3.2	4.5	2.5	1000
2017	4.8	3.5	4.2	2.8	1100
2018	4.6	3.8	4	3	1200
2019	4.4	4	4.2	3.2	1150
2020	2	2.5	5	1.5	900

In this table, each row represents data collected for a specific year, with columns for different economic indicators such as GDP growth rate, inflation rate, unemployment rate, interest rate, and stock market index. This type of data is used in econometrics to analyze and model the relationships between these variables over time.

2.10.2.1 Pros of Time Series Data

Time series data analysis has a number of benefits for comprehending and making use of temporal data. Its capacity to identify trends and patterns present in data across time is one of its main advantages. By highlighting past trends that inform present activities, these insights support strategic planning and well-informed decision-making. Time series analysis is also excellent at forecasting, enabling firms to project future values based on past trends. Because this approach is quantitative, it offers a formal framework for assessing changes and variations, allowing for a greater comprehension of complicated dynamics like cause-and-effect interactions. Time series data analysis is essential for understanding cyclic behaviours and developing fact-based forecasts in areas like scientific research and companies with seasonal changes.

2.10.2.2 Cons of Time Series Data

Analysis of time series data can provide some difficulties, though. Noise can hide important patterns and calls for careful data pretreatment because it is brought on by erratic variations and outliers. Because it can be difficult and time-consuming to guarantee data stationarity, a frequent requirement for many models, its applicability is constrained. The availability of previous data is also essential for good forecasting, perhaps impeding its application in developing industries or surroundings that change quickly. Furthermore, sophisticated models have a tendency to overfit noisy data, which might result in unreliable predictions and limited generalisation abilities. The interpretability of underlying systems may be veiled, creating a barrier in terms of actionable insights, even though time series analysis offers predictive potential.

2.10.3 Panel Data (Longitudinal Data)

A specialised subset of data called panel data, often referred to as longitudinal data, is composed of observations that are routinely made on the same people or things throughout a range of time periods. This kind of data can be represented visually as a fusion of cross-sectional and time series data, providing a multidimensional viewpoint. Cross-sectional data provides observations across various entities at a single point in time, whereas time series data collects observations across time for a single entity. However, panel data collects information about different cross sections across time, such as nations, businesses, people, or demographic groups. Panel data preserves a chronological order of observations gathered at regular intervals, much like time series data. Likewise, panel data includes a collection of observations covering numerous people or entities, just as cross-sectional data. Panel data has various advantages over other data types thanks to this special combination:

1. **Behavioural Insights:** Panel data makes it easier to model both group-level and individual-level behaviour. It captures how people or other things develop throughout time while taking numerous influences into account.
2. **Information Density:** Pure time series or cross-sectional data are not as rich or variable as panel data, which offers a more comprehensive dataset. The possibility for deriving insightful ideas and spotting patterns is increased by this richness.
3. **Detection of Statistical Effects:** Panel data has the potential to locate and measure statistical effects that might be concealed in pure time series or cross-sectional datasets. When examining relationships between the two dimensions, this is very pertinent.
4. **Reduced Estimation Biases:** Estimation biases can be introduced by combining groups into a single time series. By enabling the examination of differences both within and between groups, panel data helps to eliminate these biases.

In conclusion, panel data offers a flexible framework for comprehending the interactions and dynamic behaviour of entities through time. In analysing trends, individual variations, and statistical effects, its multidimensionality offers special advantages that can improve research and decision-making in a variety of sectors.

Table 2.4 Illustration of Panel Data

Country	Year	GDP Growth Rate (%)	Inflation Rate (%)	Unemployment Rate (%)	Interest Rate (%)	Stock Market Index
USA	2016	2.5	1.8	4.7	2	2000
USA	2017	2.3	2.1	4.4	2.2	2100
USA	2019	2.2	2.3	3.8	2.3	2150
USA	2020	1.5	1.9	4.2	1.8	1900
India	2016	7.1	5	3.5	6	3000
India	2017	6.5	4.5	3.2	5.8	3100
India	2019	6.8	4.7	3.1	6	3150
India	2020	4.2	3.5	3.8	4.5	2800

2.11 Future Trends in Data Structure Analysis

The field of data structure analysis is now experiencing rapid growth, with several significant topics projected to shape its future trajectory. The increasing incorporation of advanced technology into data collection and processing methodologies is a key determinant. Technological advancements, shown by wearables, sensors, and IoT (Internet of Things) devices, are generating substantial volumes of diverse data. To effectively integrate emerging data types, such as time-series data derived from sensors and spatial data obtained from location-based devices, it is imperative to modify existing data structures. Furthermore, there is an expectation that the integration of data structure analysis with machine learning and artificial intelligence will become more prevalent. These tools have the potential to enhance and expedite data analysis processes through the automation of tasks such as anomaly identification, pattern recognition, and predictive modelling. In addition, the management of substantial volumes of data will continue to be a significant area of focus. As data quantities see significant growth, it is imperative for data structures to possess scalability and effectiveness. In order to effectively handle large datasets, it is imperative to optimise both storage and access approaches. The significance of specialist structures such as graph databases and NoSQL databases is expected to increase in the management and analysis of intricate and interconnected data. The analysis of data structures will be subject to substantial influence from ethical considerations and concerns over privacy. In order to reconcile the need for comprehensive analysis and the protection of individual privacy rights, it will be imperative to employ novel techniques in data anonymization, encryption, and access control.

Technological advancements are radically altering the landscape of data collection and analysis. The proliferation of interconnected devices and sensors has led to the generation of real-time data streams, hence highlighting the need for efficient methodologies in data processing and storage. Consequently, there has been a surge in the adoption of distributed computing frameworks such as Apache Hadoop and Spark. The advancement of edge computing is facilitating the analysis of data in closer proximity to its source, hence reducing latency and enhancing the ability to make real-time decisions. Contemporary data ecosystems are employing nascent tools and approaches such as data lakes and data warehouses. These tools enable organisations to collect, store, and analyse extensive and diverse datasets. Cloud computing platforms play a crucial role in offering scalable infrastructure for the processing and storage of data, thereby freeing enterprises from the constraints imposed by physical hardware located on their premises. Furthermore, advancements in the field of text analysis and natural language processing (NLP) are facilitating the extraction of valuable insights from unstructured textual data. This is of utmost importance for sentiment analysis, social media monitoring, and customer feedback analysis. The trajectory of data-driven decision-making is being influenced by the interdependent relationship between technology and the analysis of data structures. The convergence of advanced technology, innovative tools, and ethical considerations will shape the methodologies employed in the collection, storage, analysis, and utilisation of data within diverse corporate contexts.

2.12 Recent Data Structure Trends

The field of business research has witnessed significant advancements in data structuring techniques. These trends reflect the increasing complexity and volume of data being generated by businesses, as well as the need for more efficient ways to organize, store, and analyse this data. Some of the key trends in data structure that researchers and practitioners should be aware of include:

a) **Big Data and Data Lakes:** With the proliferation of digital technologies, businesses are now collecting vast amounts of data from various sources such as social media, sensors, and transactions. To manage this influx of data, many organizations are adopting big data technologies and storing data in data lakes—a centralized repository that allows for the storage of structured, semi-structured, and unstructured data at scale.

b) **NoSQL Databases:** Traditional relational databases are often ill-suited for handling the variety and volume of data generated by modern businesses. As a result, many organizations are turning to NoSQL databases, which are designed to handle large volumes of data in a flexible, schema-less manner. NoSQL databases can be particularly useful for storing unstructured data such as text, images, and videos.

c) **Data Virtualization:** Data virtualization is a technique that allows organizations to access and manipulate data from multiple sources without the need to physically move or replicate the data. This can be particularly useful for businesses that have data spread across multiple systems and need to integrate this data for analysis purposes.

d) **Graph Databases:** Graph databases are designed to represent and store relationships between data points. This makes them particularly well-suited for applications such as social network analysis, fraud detection, and recommendation engines, where the relationships between data points are as important as the data points themselves.

e) **Time-Series Databases:** Time-series databases are optimized for storing and querying time-stamped data, such as stock prices, sensor data, and IoT data. These databases are designed to efficiently handle large volumes of data points and are often used in applications that require real-time analytics.

f) **Blockchain:** While initially popularized by cryptocurrencies, blockchain technology has potential applications beyond finance. In business research, blockchain can be used to create secure, transparent, and tamper-proof records of transactions and other data, making it particularly useful for applications such as supply chain management and intellectual property protection.

g) **Cloud-Based Data Warehousing:** Cloud-based data warehousing solutions are becoming increasingly popular among businesses due to their scalability, flexibility, and cost-effectiveness. These solutions allow organizations to store and analyze large volumes of data without the need to invest in expensive on-premise infrastructure.

h) **Data Governance and Privacy:** As businesses collect and store more data, there is a growing concern around data governance and privacy. Organizations are increasingly implementing robust data governance policies and practices to ensure that data is managed and protected in accordance with legal and regulatory requirements.

2.13 Summary

This chapter provides a comprehensive overview of scales in research, covering nominal, ordinal, interval, and ratio scales, and their respective characteristics, uses, properties, and tests for adequacy. It emphasizes the importance of selecting the right scale for different data and research needs, while also recognizing the challenges and limitations associated with each scale type. Practical examples and calculations are included to illustrate the application of these scales in research, helping readers understand and apply them effectively. Additionally, the chapter discusses different types of data structures, highlighting their importance in modern research and providing insights into organizing, storing, and analysing data for improved research outcomes. In conclusion, recent trends in data structure reflect the growing complexity and volume of data being generated by businesses, as well as the need for more efficient ways to organize, store, and analyse this data. By staying abreast of these trends, researchers and practitioners can ensure that their data structuring techniques are aligned with the evolving needs of their organizations.

2.14 Self-Assessment

1. Explain the concept of a nominal scale and provide an example of its application in business research.
2. Compare and contrast ordinal and interval scales, highlighting their key differences and similarities.
3. Discuss the importance of selecting the right scale for different types of data in research.
4. Describe the properties of a ratio scale and explain why it is considered the most informative scale type.
5. What are some common tests for assessing the adequacy of a scale in research?
6. How do recent advancements in scale development impact the field of business research?
7. Provide a practical example of using a nominal scale in a research study, including the data collection and analysis process.
8. Explain how understanding different types of data structures can improve the quality of research outcomes.
9. Discuss the challenges and limitations associated with using ordinal scales in research, and suggest strategies to overcome them.
10. How can researchers stay updated with recent advancements in scale development and apply them in their research projects?

Unit-III

Collection and Presentation of Data

3.1 Learning Objectives

After completing this chapter, you should be able to:

i) Define primary data and secondary data and their applications in various contexts;

ii) Identify and briefly explain the various methods used for collecting primary data;

iii) Recognize the significance of secondary data and its advantages and disadvantages;

iv) Create a pie chart, histogram, bar chart, column chart, and line chart using appropriate data sets;

v) Evaluate the advantages and limitations of each data presentation method.

3.2 Introduction

Presently, the role of data has transcended from mere numbers and figures to a powerful driving force behind informed decision-making, efficient problem-solving, and insightful analysis. This chapter delves into the crucial aspects of data collection and presentation and it will help in unraveling their significance across a spectrum of professional fields. The chapter will help to explore the fundamental concepts of data collection and its differentiation into primary and secondary categories. We will learn about the methods employed to gather firsthand data and from existing sources. Furthermore, we will delve into the realm of data presentation, understanding how various visualization techniques such as pie charts, histograms, bar and column charts, and line charts transform data into digestible and meaningful representations

3.3 Importance of Data Collection and Presentation

In every sphere of human endeavour, be it business, healthcare, education, or scientific research, the importance of data collection and presentation cannot be overstated. Data serves as the foundational building block upon which strategies, solutions, and innovations are constructed. Whether it's a corporation strategizing its next market move, a healthcare professional determining treatment options, an educator tailoring instruction, or a researcher uncovering patterns in the cosmos, data stands at the epicenter of progress.

3.4 Role of Data in Decision-Making

Data, in its processed forms, forms the bedrock of decision-making. Every choice made in the professional world relies on information, and data provides the empirical foundation for these choices. Data helps professionals in taking decisions effectively as it is based on tangible evidence rather than intuition or guesswork. By analyzing trends, correlations, and historical patterns, professionals can predict outcomes, forecast trends, and mitigate risks with a higher degree of confidence. Data plays a significant role in the decision-making process and these are as follows:

i) **Facilitating Problem-Solving:** Nowadays, data-driven problem-solving has become an invaluable asset due to its efficient data handling and quick recommendation capacity. The ability to collect, analyze, and interpret data enables professionals to deconstruct complex problems, identify root causes, and devise targeted solutions. This process is not limited to reactive troubleshooting; it also encompasses proactive anticipation of issues, allowing for preventive measures to be put in place.

ii) **Enabling In-depth Analysis:** Data collection is the primary step in the journey towards insightful analysis. By gathering diverse data points, professionals can delve deep into their domains, uncovering patterns, anomalies, and relationships that might otherwise remain hidden. Through systematic analysis, professionals can extract actionable insights, drive innovation, optimize processes, and refine strategies.

3.5 Primary Data and Its Collection Techniques

3.5.1 Definition of Primary Data

Primary data stands as the bedrock of empirical research and analysis, representing the raw, unprocessed information collected firsthand from its original source. Primary data serves as the closest and most direct form of information acquisition and provides a direct insight into the subject of study. This section sheds light on the nature and significance of primary data, highlighting its role in generating accurate and reliable insights.

At its core, primary data encompasses information gathered specifically for the purpose of the research or analysis at hand. This could involve conducting surveys, interviews, experiments, or observations to collect data that directly pertains to the objectives of the study. Primary data is akin to collect a firsthand account of an event, phenomenon, or individual, rather than relying on pre-existing records or summaries.

3.5.2 Importance of Primary Data

The significance of primary data lies in its authenticity and relevance. As primary data is collected directly from the source, it reflects the most accurate representation of the subject under study. Researchers can tailor their data collection methods to suit the specific requirements of their research questions, ensuring that they capture nuanced information that might be missed when relying solely on secondary sources.

3.5.3 Methods of Collecting Primary Data

Primary data collection methods are diverse and adaptable, allowing researchers to employ strategies that align with their research goals and the nature of their subjects. Some common methods are as follows:

i) **Surveys:** Questionnaires and interviews are widely used to collect data from a sample population. Surveys enable researchers to gather standardized information from a large number of respondents.

ii) **Observations:** Observational studies involve the systematic recording of behaviours, events, or phenomena as they occur naturally. This method is often used in social sciences and naturalistic research.

iii) **Experiments:** Controlled experiments involve manipulating variables to observe their effects on outcomes. This method is prevalent in scientific research and allows for the establishment of cause-and-effect relationships.

iv) **Case Studies:** In-depth investigations of a single individual, group, or entity provide rich and detailed insights. Case studies are common in fields such as psychology, sociology, and healthcare.

3.5.4 Advantages and Disadvantages of Primary Data

Primary data has certain advantages and disadvantages although it offers unique insights and accuracy. The advantages of primary data are as follows:

i) **Relevance and Specificity:** Primary data is tailored to the research objectives, ensuring that the collected information directly addresses the research questions. This level of specificity enhances the relevance of the data.

ii) **Accuracy and Authenticity:** Since primary data is collected firsthand, it is more likely to be accurate and authentic. Researchers have control over the data collection process, minimizing errors, and inaccuracies.

iii) **Control over Variables:** Researchers have the ability to manipulate variables and experimental conditions in controlled experiments, allowing for the establishment of cause-and-effect relationships.

iv) **Freshness of Data:** Primary data is collected in real time, making it current and reflective of the current state of affairs. This is especially valuable when researching rapidly changing phenomena.

v) **Depth of Insights:** Primary data collection methods such as interviews and observations provide rich, qualitative insights that might not be captured by secondary data sources.

Primary Data has also Certain Disadvantages and These are as follows

i) **Time and Resource Intensive:** Collecting primary data can be time-consuming and requires significant resources. Designing surveys, conducting interviews, and performing experiments demands careful planning and execution.

ii) **Potential Bias:** Researcher bias can inadvertently influence the data collection process and subsequent analysis. Efforts must be made to minimize subjectivity.

iii) **Limited Sample Size:** Depending on the research scope and resources available, the sample size for primary data collection might be smaller compared to secondary data sources.

iv) **Generalizability Concerns:** Findings based on a specific sample might not always generalize to the broader population. Researchers must be cautious when extrapolating results.

v) **Complexity of Analysis:** Analyzing primary data, particularly qualitative data, can be complex and require specialized skills and tools.

vi) **Ethical Considerations:** Collecting primary data involving human subjects requires careful consideration of ethical guidelines and obtaining informed consent.

vii) **Potential Incomplete Data:** There might be cases where participants do not provide complete or accurate information, affecting the quality of the collected data.

3.6 Secondary Data and Its Collection Techniques

3.6.1 Definition of Secondary Data

Secondary data plays a vital role in research and analysis by providing information that has already been collected by others for a different purpose while primary data offers a direct and personalized perspective. Secondary data is data that has been gathered, compiled, and documented by researchers, institutions, or organizations for their own research objectives. It is not collected specifically for the study at hand but can be repurposed to address new research questions. Simply, a researcher collected data for a specific project and made it available for others to use. When another scholar uses this particular data for a different project, it is referred to as secondary data. Secondary data encompasses a wide range of information, including historical records, official documents, scholarly articles, and datasets.

3.6.2 Sources of Secondary Data

Secondary data can be obtained from various sources, each contributing to a diverse pool of information. Some common sources include:

i) **Government Agencies:** Government bodies collect an array of data related to demographics, economics, public health, and more. Census data, economic indicators, and crime statistics are examples of valuable secondary data available through government agencies.

ii) **Research Institutions:** Academic institutions, research centres, and think tanks conduct studies and experiments, generating a wealth of data. Research reports, surveys, and academic papers can serve as rich sources of secondary data.

iii) **Published Reports and Studies:** Published reports, whether from government bodies or private organizations, offer insightful data that has undergone rigorous analysis. These reports often cover topics ranging from market trends to environmental impact.

iv) **Online Databases:** The digital era has ushered in an age of easily accessible online databases. These repositories house a wide range of data, from scientific research to social trends, providing researchers with a treasure trove of secondary data.

3.6.3 Advantages and Disadvantages of Secondary Data

Secondary data has several advantages that make it a valuable resource in research and analysis. The advantages are as follows:

i) **Time and Cost Efficiency:** Since the data already exists, researchers can save time and resources that would otherwise be spent on data collection.

ii) **Historical Insights:** Historical data provides a longitudinal view of trends and changes over time, aiding in the understanding of patterns and developments.

iii) **Comparative Analysis:** Secondary data allows for comparisons across different time periods, regions, or demographics, enriching the depth of analysis.

iv) **Large Sample Sizes:** Data collected by established institutions often comes from large sample sizes, enhancing the statistical robustness of analyses.

However, secondary data has also certain disadvantages that researchers should be aware of when considering its use in their studies. These disadvantages are as follows:

i) **Lack of Relevance:** Secondary data might not precisely align with the specific research questions or objectives of a new study. It was originally collected for a different purpose, potentially resulting in irrelevant or incomplete information.

ii) **Quality and Accuracy:** The quality and accuracy of secondary data can vary widely. Researchers have limited control over the data collection process, leading to potential errors or biases in the original source.

iii) **Limited Context:** Secondary data might lack the contextual information and nuances that are crucial for a comprehensive understanding of the research topic. Without proper context, data interpretation can be challenging.

iv) **Inconsistencies and Incompatibilities:** Data collected by different sources using various methodologies might be inconsistent or incompatible when combined. Researchers need to carefully assess and clean the data before analysis.

v) **Limited Control over Variables:** Researchers have little or no control over how variables were measured and collected in the original study. This lack of control can impact the validity of comparisons or analyses.

vi) **Data Availability:** Not all desired information might be available in secondary data sources. This can lead to gaps in the data, preventing researchers from fully addressing their research questions.

vii) **Ethical Concerns:** When using secondary data involving human subjects, ethical considerations surrounding data privacy and consent might be less transparent compared to primary data collection.

viii) **Limited Flexibility:** Researchers are confined to the data variables and formats provided by the original sources. This limitation might prevent a deeper exploration of certain aspects of the research question.

3.7 Frequency Distribution

Data, in its raw form, can often be overwhelming and difficult to interpret. This is where the concept of frequency distribution steps in as a powerful tool to organize, summarize, and uncover patterns within data. Frequency distribution is a methodical arrangement of data into categories or intervals along with the corresponding frequencies (f) of occurrences. By grouping data points into intervals, researchers can transform a long list of individual values into a structured format that highlights the distribution of values across the range. This organized presentation enables analysts to distinguish trends, variations, and concentrations within the data. Frequency distribution is primarily presented as charts or frequency tables. Frequency distribution can present in an actual number of observations within specific ranges or the proportion (%) they represent. The representation of frequency in percentage is referred to as a ***relative frequency*** distribution. Frequency distribution can be used for both categorical and numeric variables. For continuous (numeric) variables, class interval creation is required, which will be described shortly.

Advantages of Frequency Distribution

Frequency distribution offers several advantages that contribute to the clarity and interpretation of data:

i) **Simplification:** Complex data sets are simplified into manageable groups, making it easier to identify trends and outliers.

ii) **Visualization:** Frequency distribution tables can be complemented by visual representations like histograms, making data patterns more accessible to a wider audience.

iii) **Identification of Patterns:** By displaying the frequency of values within intervals, researchers can identify concentrations, gaps, or anomalies in the data distribution.

iv) **Comparison:** Frequency distributions allow for easy comparison of different data sets or segments of a larger dataset.

v) **Decision-Making:** Insights derived from frequency distributions assist in informed decision-making by providing a clearer understanding of data behaviour.

3.7.1 General Steps in Creating a Frequency Distribution

Creating a frequency distribution involves a structured sequence of steps that guide the transformation of raw data into an insightful frequency table. The steps are as follows:

Step 1 Determine the Range and Class Intervals

Starting by identifying the range of values present in the dataset. This helps in determining the overall span of the frequency distribution. Next, decide on appropriate class intervals—ranges of values within which data will be grouped. The choice of class intervals should reflect the distribution's complexity and the desired level of detail.

Step 2: Tally the Data into Respective Intervals

Organize the data by tallying each value into its corresponding class interval. This step creates a preliminary frequency distribution.

Step 3: Calculate the Frequency for Each Interval

Count the number of tallies within each class interval to determine the frequency of occurrences. This reveals how many data points fall within each interval.

Step 4: Create a Frequency Table

Compile the calculated frequencies into a tabular format known as a frequency table. This table presents class intervals and their respective frequencies, offering a visual summary of the data's distribution.

There are primarily two types of frequency distribution and those are ungrouped and grouped frequency distribution. The below examples will help you to construct both grouped and ungrouped data frequency tables.

3.7.2: Example 1: Creating a Frequency Distribution Table for Ungroup Data

A car company surveyed 20 households to know how many cars they registered. The number of registered cars is as follows:

0, 1, 4, 3, 2, 1, 2, 1, 1, 2, 3, 2, 2, 1, 1, 1, 2, 1, 0, 0

Use the above steps to construct a frequency table. First, we need to divide the results into intervals. In this case, the intervals will be the number of households with no cars (0), one car (1), two cars (2), three cars (3), and so on. Second, we need to proceed to count the numbers within each interval. Construct a table featuring distinct columns for the interval values (indicating the number of cars per household), the tally marks for each interval, and the frequency of occurrences within each interval. Label these columns as "Number of Cars," "Tally," and "Frequency," respectively.

Number of Cars	Tally	Frequency
0		
1		
2		
3		
4		

Third, review the data list from left to right and position a tally mark in the corresponding row. As an illustration, if the first data point is 1, insert a tally mark in the row corresponding to the interval value of 1 in the "Number of Cars" column. Subsequently, if the next data point is 2, insert a tally mark in the row for the interval value 2, and continue this process. After accumulating five tally marks in a row, replace them with a tally line across the preceding four marks. This practice streamlines the process of calculating the final frequencies. For example, if we have six households under no car group then the tally will look like this (*~~IIII~~ I*). Fourth, sum up the tally marks in each row and document these totals in the "Frequency" column, which serves as the conclusive record of frequencies for each interval.

Number of Cars	Tally	Frequency
0	*III*	3
1	*~~IIII~~ III*	8
2	*~~IIII~~ I*	6
3	*II*	2
4	*I*	1
	TOTAL	20

From the table, it is noticed that 3 households have no cars, 8 households have one car, 6 households have two cars, 2 households have three cars, and 1 household has 4 cars.

3.7.3 Example 2: Creating a Frequency Distribution for Grouped Data

A car company surveyed 20 households to know the age group of the people who drive the car. The ages are as follows:

40, 43, 56, 34, 23, 56, 42, 31, 49, 59, 22, 43, 45, 48, 32, 44, 37, 28, 29, 32

In this case, we observed that the value range is large (the low value is 22 and the high value is 59). So, we will create a class to make the data more presentable. However, we will follow the same steps to determine the frequency of each class. Here, we have created a class with interval 10 and have presented the frequency distribution in the below table:

Class (age group)	Tally	Frequency
20-29	*IIII*	4
30-39	*~~IIII~~*	5
40-49	*~~IIII~~ III*	8
50-59	*III*	3
Total		20

3.8 Relative Frequency Distribution

Relative frequency distribution is an important concept in statistics that deals with how often different values occur in a dataset, expressed as a fraction or percentage of the total number of occurrences. This distribution is particularly useful for understanding the proportion of each category within a dataset, making it easier to compare different categories or datasets of different sizes.

To calculate the relative frequency of any given data value, you first need to divide its frequency (the number of times it appears in the dataset) by the total frequency of all data values in the set. This calculation can be represented by the formula: Relative Frequency = f/n, where 'f' is the frequency of the specific data value and 'n' is the total frequency of all data values.

Example: We will use the previously developed grouped data for calculating the relative frequency.

Class (age group)	Tally	Frequency (f)	Relative Frequency (f/n)
20-29	*IIII*	4	4/20=0.2
30-39	*~~IIII~~*	5	5/20=0.25
40-49	*~~IIII~~ III*	8	8/20=0.4
50-59	*III*	3	3/20=0.15
Total		20 (n)	1

3.9 Cumulative Frequency Distribution

Cumulative frequency distribution is a statistical tool used to understand the cumulative effect of frequencies over a data set. It's particularly useful when the data has a natural order, allowing for meaningful analysis of the running total of frequencies.

To define it simply, a cumulative frequency distribution is the sum of the frequencies for a value in a dataset and all the values that precede it. For instance, if you have frequencies for different categories or class intervals, the cumulative frequency for a particular class interval would be the sum of its frequency and all the frequencies of the intervals before it.

There are two types of cumulative frequency distributions: "less than" and "greater than" types. The "less than" type shows the running total of frequencies for values less than or equal to a certain value, while the "greater than" type does the opposite, showing frequencies for values greater than or equal to a specific value.

Graphical representation of cumulative frequency is also possible through cumulative frequency curves or ogives. These are useful for visualizing the data distribution and can be particularly helpful in determining the median or other statistical measures.

Example: We will use the previously developed grouped data for calculating the cumulative frequency.

Class (age group)	Frequency (f)	Cumulative Frequency	
		Less than	More than
20-29	4	4	20
30-39	5	9	16
40-49	8	17	11
50-59	3	20	3
Total	20 (n)		

3.10 Presentation of Data

3.10.1 Importance of Effective Data Presentation

At present, data alone is not enough; it must be presented in a manner that is comprehensible, impactful, and easy to interpret. This section delves into the crucial role of data presentation, particularly through the lens of visualization, in enhancing communication, facilitating decision-making, and enabling efficient understanding of complex information.

Data visualization serves as a powerful tool to transform numbers and statistics into visual forms that transcend language barriers and cognitive limitations. By employing graphs, charts, and other graphical representations, data visualization takes advantage of our natural visual perception, making it easier to identify trends, patterns, and outliers in datasets. Effective data presentation is a cornerstone of effective communication, enabling professionals to share insights with colleagues, clients, and stakeholders.

3.10.2 Types of Data Presentation

Data is primarily presented in three different forms and those are as follows:

i) Textual Presentation

ii) Tabular Presentation

iii) Graphical Presentation

3.10.3 Textual Presentation

This is the simplest and most direct method of presenting data. In textual presentation, data is described within the text, usually in paragraph format. It's particularly useful for supplementing qualitative statements with quantitative data. For instance, in a report discussing an earthquake's impact, you might find a sentence like, "The 2002 earthquake resulted in approximately 10,000 fatalities." This method is best used when the data set is not extensive and the key facts or figures can be easily described in a few sentences.

3.10.4 Tabular Presentation

When it comes to organizing and presenting large amounts of data in an easily understandable format, tables are extremely effective. A table arranges data in rows and columns, making it easier to compare different data sets and to perform statistical analysis like calculating measures of central tendency. Key components of a good table include a clear title, headnotes, stubs (row titles), captions (column titles), and sometimes footnotes and source citations. Tables are particularly advantageous because they can represent a large amount of data in a compact form, facilitating easier comparison and analysis.

Example 1: One-way Table

Table No. 1: Department-wise distribution of university students (n=100)

Department	No. of students
Science	20
Arts	50
Commerce	20
Management	10

Footnote

Source

Example 2: Two-way Table

Table No. 2 Classification of university students (department and sex-wise) (n=100)

Department	No. of students		
	Male	Female	Total
Science	10	10	20
Arts	20	30	50
Commerce	15	5	20
Management	6	4	10
Total	51	49	100

Example 3 Three-way Table

Department	Male			Female			Total		
	Hindu	Non-Hindu	Total	Hindu	Non-Hindu	Total	Hindu	Non-Hindu	Total
Science	6	4	10	8	2	10	14	6	20
Arts	14	6	20	25	5	30	39	11	50
Commerce	8	7	15	3	2	5	11	9	20
Management	5	1	6	3	1	4	8	2	10
Total	33	18	51	39	10	49	72	28	100

3.10.5 Graphical Presentation

Graphs and charts are powerful tools for visualizing data, especially when dealing with complex data sets or when you need to show trends, patterns, or relationships within the data. Common types of graphical presentations include bar charts, line graphs, pie charts, and histograms. These visuals are more engaging and can make complex data more accessible to a general audience. The choice of graph or chart largely depends on the type of data you're presenting and the message you want to convey. For instance, line graphs are excellent for showing changes over time, while pie charts are useful for illustrating proportions.

Graphical presentation can be done in various ways. However, some of the widely used diagrams are as follows:

i) Line chart or Line graph
ii) Bar chart or Bar graph
iii) Pie chart or Pie graph
iv) Histogram
v) Frequency polygon
vi) Cumulative frequency polygon or ogive

3.10.5.1 Line Chart

A line chart is a powerful visualization tool that is particularly useful for displaying trends and patterns over time or across ordered categories. It excels in showing the progression of data points and revealing changes, fluctuations, and relationships within datasets. Line charts are especially valuable when you want to emphasize the overall trend and how individual data points contribute to that trend.

When to Use

Line charts are well-suited for a range of scenarios:

i) Time Series Data: Displaying trends, changes, and patterns over time, such as stock prices, temperature variations, or sales figures.

ii) Comparing Trends: Comparing trends across multiple variables or categories, demonstrating how different variables evolve over the same time period.

iii) Identifying Patterns: Identifying cyclic patterns, seasonality, or irregularities in data sequences.

Constructing a Line Chart

Constructing a line chart involves the following steps:

i) **Collect and Arrange Data:** Gather data that includes multiple data points with a temporal or ordinal relationship, such as time periods or categories with a logical order.

ii) **Choose Axes:** Assign the horizontal (x) axis to the time periods or ordered categories and the vertical (y) axis to the variable being measured.

iii) **Plot Data Points:** For each time period or category, plot the corresponding data point on the chart using a dot or marker.

iv) **Connect Data Points:** Draw lines connecting the data points on the chart. The lines visually highlight the trends and connections between consecutive data points. A line moving upwards indicates a positive trend or increase in the variable being measured over time. However, a line moving downwards indicates a negative trend or decrease. A flat line suggests stability or no significant change over the observed periods. A curved line can indicate nonlinear trends or patterns.

Example 1: The monthly production of the number of cars of a certain company for the first six months of the year 2022 is given below. Represent the data given below by a line chart.

Month	**January**	**February**	**March**	**April**	**May**	**June**
No. of Cars	200	210	220	250	260	280

Solution

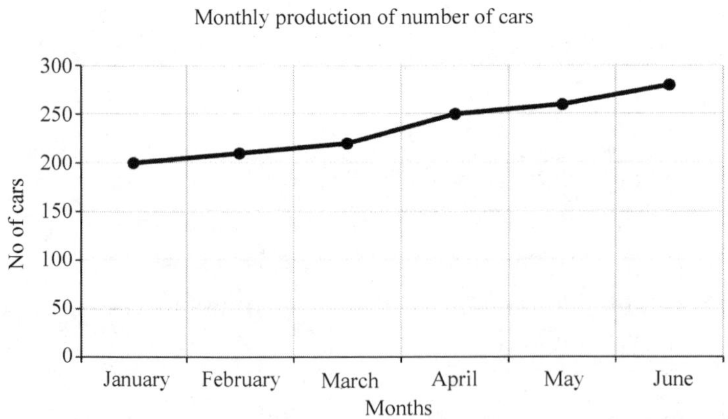

Example 2: The monthly sales of the number of cars in two cities of a certain company for the year 2023 are given below. Represent the data using a line graph.

Months	Jan	Feb	Mar	Apr	May	June	July	Aug	Sept	Oct	Nov	Dec
City 1	4	6	7	10	15	13	12	11	9	7	5	3
City 2	9	10	12	4	2	1	1	4	6	8	10	12

Solution

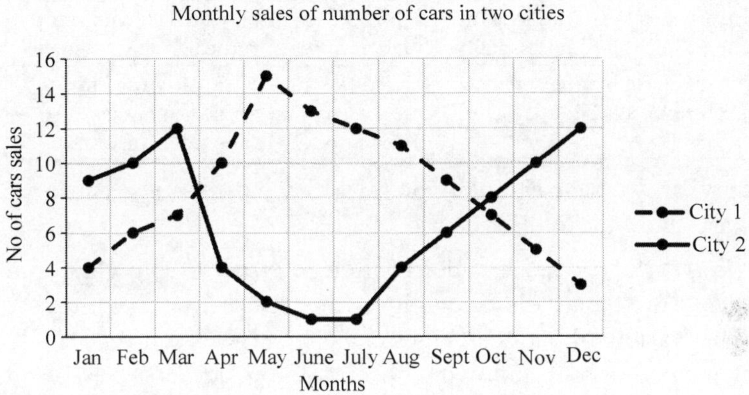

3.10.5.2 Bar Chart

Bar charts and column charts are versatile and widely used data presentation methods that effectively display categorical data and comparisons. Both charts represent data using rectangular bars, but they differ in their orientation and application. Based on the orientation, bar chart is two types.

i) **Horizontal Bar Chart:** In a horizontal bar chart, the bars are horizontal, and the lengths of the bars correspond to the values they represent. Each bar represents a separate category, and the chart is often used to compare the sizes of different categories.

ii) **Vertical Bar Chart/Column Chart:** In a vertical bar or column chart, the bars are vertical, and the heights of the bars correspond to the values they represent. Similar to bar charts, each column represents a separate category, and the chart aids in comparing the sizes of different categories.

Differences between Horizontal and Vertical Bar Charts

	Horizontal Bar Chart	**Vertical Bar/Column Chart**
Orientation	Horizontal bars	Vertical bars
Comparing categories	Used to compare categories on the y-axis	Used to compare categories on the x-axis
Labelling	Effective when labels are long or need to be positioned horizontally	Suitable for displaying data with short labels or when vertical space is limited
Application	It is mainly used to represent data categorized by attributes	Vertical bars are primarily used to represent time series data or data categorised by values of a variable

For each of these types, we have again divided it into two categories. These are Grouped Bar Charts and Stacked Bar Charts.

Grouped Bar Chart: A grouped bar chart, often referred to as a **clustered** bar graph, is utilized to display discrete values for multiple items within the same category. This variety of bar charts aggregates the total occurrences into a single column for each group, allowing for a side-by-side comparison of different data sets. Each series within the dataset is typically represented by a distinct colour to aid in differentiation. Clustered bar charts can be oriented both vertically and horizontally, facilitating comparison across different categories.

Stacked Bar Chart: Stacked Bar Chart also alternatively known as a **composite** bar graph, the composite bar chart breaks down a total quantity into its component segments. This format of bar chart distinguishes each segment by colour, making it straightforward to distinguish between categories. To accurately convey the composition of the bars, detailed labels are necessary. In a composite bar chart, the entirety of each bar signifies the complete data set, while the individual sections of the bar illustrate the proportional parts of the total.

Example 1: In a school of 109 students, the percentage of monthly savings by an employee is given in the below table. Create a vertical bar chart using the data.

Savings (%)	10	20	30	40	50	60
No of student	30	40	20	10	6	3

Solution [Vertical Bar Chart]

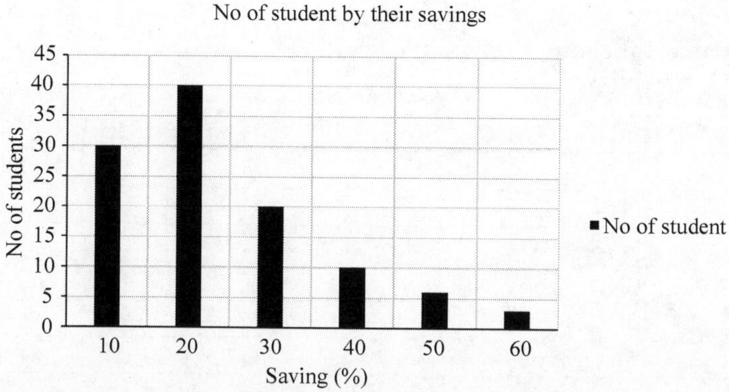

Example 2: The monthly sales of the number of cars in nine metropolitan cities of a certain company for the year 2023 are given below. Represent the data using a bar graph.

Cities	**Delhi**	**Kolkata**	**Mumbai**	**Ahmedabad**	**Chennai**	**Surat**	**Hyderabad**	**Pune**	**Bangalore**
No. of Car Sales	12	6	7	10	15	13	12	11	8

Solution [Horizontal Bar Chart]

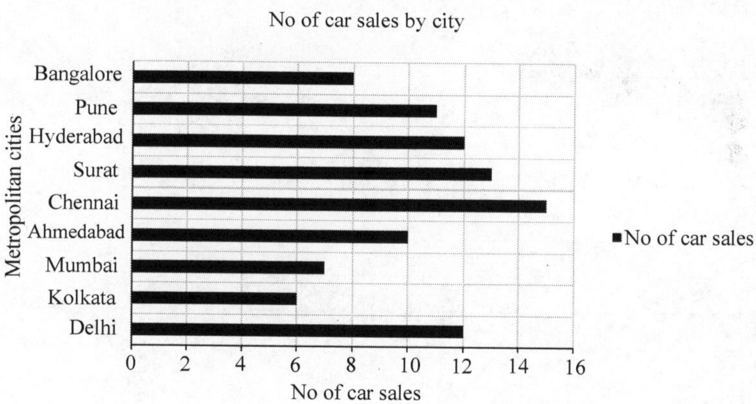

Example 3: A car company manufactures 3 different models of car. The sale for 6 months is shown in the table. Prepare a bar graph.

Month	No. of car sales		
	Model 1	Model 2	Model 3
Jan	13	8	15
Feb	14	12	13
Mar	17	14	12
Apr	18	14	8
May	12	10	15
Jun	8	6	17

Solution [Grouped/Clustered Bar Chart]

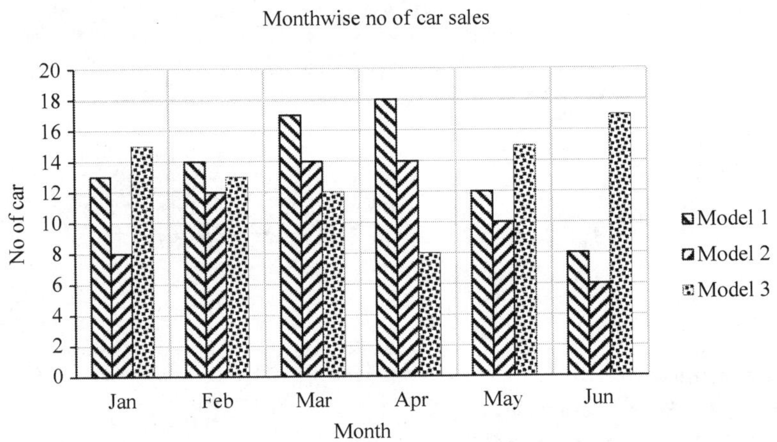

Example 4: Prepare a bar chart from the following year-wise agricultural and non-agricultural data.

Year	Agricultural	Non-agricultural
2010	12	10
2011	14	11
2012	15	12
2013	16	13
2014	19	11
2015	18	14
2016	21	16
2017	23	17
2018	26	18

Solution [Stacked Bar Chart]

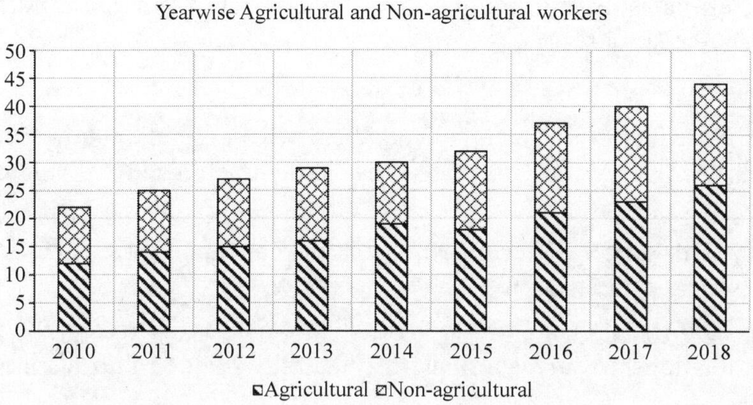

3.10.5.3 Pie Chart

A pie chart is a circular graph divided into slices that represent different parts of a circle. Each slice's size is proportional to the quantity it represents within the entire dataset. Pie charts are effective for showcasing relative proportions and the composition of a categorical variable. They are best used when displaying data with distinct categories and clear differences in magnitude.

Creating a Pie Chart

i) **Choose Relevant Data:** Select a dataset with discrete categories that you want to represent in the pie chart.

ii) **Calculate Percentages:** Calculate the percentage of each category's value in relation to the whole dataset.

iii) **Determine Slices:** Convert the percentages into degrees of a circle. Since a circle is 360 degrees, each slice's angle will correspond to the percentage it represents.

iv) **Draw the Chart:** Using graphing software or tools, draw the pie chart with each slice's angle proportional to the calculated degrees.

v) **Label the Slices:** Label each slice with the corresponding category and percentage.

Advantages and Disadvantages

Advantages

i) **Clear Representation:** Pie charts vividly display the proportion of different categories, making it easy to grasp relative sizes.

ii) **Visual Impact:** Their visual appeal captures attention and aids quick comprehension.

iii) Simple Comparison: It's straightforward to compare the relative sizes of different categories.

Disadvantages

i) Limited Categories: Pie charts work best with a small number of categories. Too many slices can make the chart cluttered and difficult to interpret.

ii) Difficulty with Small Differences: Distinguishing small differences in slice sizes can be challenging.

iii) Complex Data: Pie charts are less effective when trying to represent complex relationships or data that isn't naturally divided into distinct categories.

Example: Prepare a pie chart based on the following country's rice export data for the year 2024:

Country	India	Thailand	Vietnam	Pakistan
Export (B$)	107.7	39.6	25.2	23.5

Solution

Country	Export (B$)	Percent of total	Angle at the center of the pie chart [Percent*3.6]
India	107.7	[(107.7÷196)*100]=54.95	197.82
Thailand	39.6	[(39.6÷196)*100]=20.20	72.73
Vietnam	25.2	[(25.2÷196)*100]=12.86	46.29
Pakistan	23.5	[(23.5÷196)*100]=11.99	43.16
Total	196	100	360

Pie chart Showing Pricipal Exporting Counties of Rice (2024)

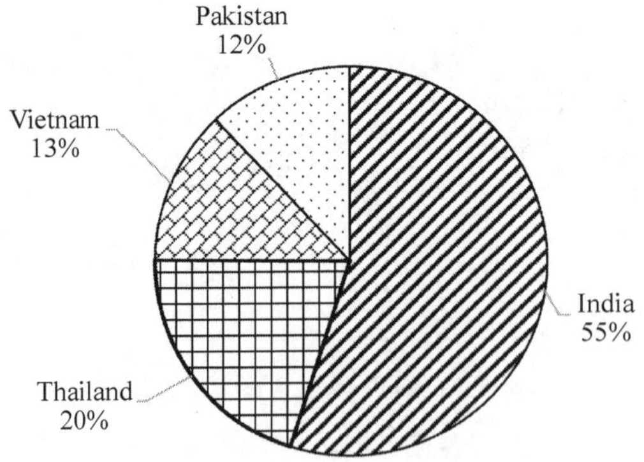

3.10.5.4 Histogram

A histogram displays the frequency distribution of a continuous variable, showing how often different values occur within specified intervals. It is especially useful when you want to understand the distribution of data, identify central tendencies, assess the spread of values, and locate potential outliers. Histograms provide insights into the underlying structure of the data, making them an essential tool for exploratory data analysis.

Steps to Construct a Histogram

i) **Choose the Number of Bins:** Decide on the number of intervals, or bins, you want to use in the histogram. Too few bins might obscure important details, while too many can result in excessive detail and noise.

ii) **Determine Bin Width:** Calculate the width of each bin by dividing the range of data values by the number of bins. This ensures that each bin covers an equal interval of the data range.

iii) **Group Data into Bins:** Organize data points into the respective bins based on their values. A data point falls into a bin if it falls within the range defined by that bin.

iv) **Count Frequencies:** Count the number of data points that fall into each bin. This count is the frequency of data points within that bin.

v) **Plot the Histogram:** On the horizontal axis, represent the range of data values, and on the vertical axis, represent the frequency or count. Create rectangles (bars) above each bin, where the height of the rectangle corresponds to the frequency of data points within that bin.

Advantages and Disadvantages

Advantages

i) **Visual Clarity:** Histograms provide a clear visualization of data distribution, making it easy to identify trends and patterns.

ii) **Distribution Insights:** Histograms help identify modes, central tendencies (mean, median, mode), and spread (range, variance).

iii) **Data Clustering:** Clusters and gaps in the histogram highlight distinct groups or concentrations within the data.

Disadvantages

i) **Sensitivity to Binning:** The choice of bin width and count can influence the appearance of the histogram and the insights derived from it.

ii) **Limited for Small Samples:** Histograms might not be as informative when working with small datasets, as the distribution might not be represented.

iii) **Subjectivity:** Interpretation of histograms can be subjective, especially when binning choices affect the visual appearance.

Example: Prepare a Histogram for the following data:

Weight (kg)	30-40	40-50	50-60	60-70	70-80	80-90
No of student	200	320	500	540	400	100

Solution

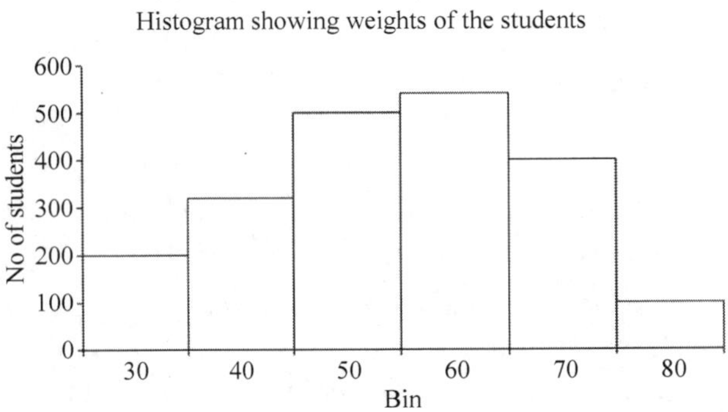

3.10.5.5 Frequency Polygon

A frequency polygon is a graphical tool used in statistics to represent and analyze the distribution of data. It serves a similar purpose as histograms but is particularly advantageous for comparing different data sets. Frequency polygons utilize line graphs to depict quantitative data. They are most beneficial when representing cumulative frequency distributions or when comparing the shapes of different distributions. To construct a frequency polygon, you plot the midpoints of the class intervals on the x-axis against the frequencies on the y-axis and then connect these points with straight lines.

Example: Construct a frequency polygon for the following data:

Class boundary	No of student
40-49	8
50-59	12

60-69	16
70-79	9
80-89	6

Solution

Class limits	Class boundaries	Mid-value	No of student
30-39	29.9-39.9	34.9	0
40-49	39.9-49.9	44.9	8
50-59	49.9-59.5	54.9	12
60-69	59.5-69.5	64.9	16
70-79	69.5-79.5	74.9	9
80-89	79.5-89.5	84.9	6
90-99	89.5-99.5	94.9	0

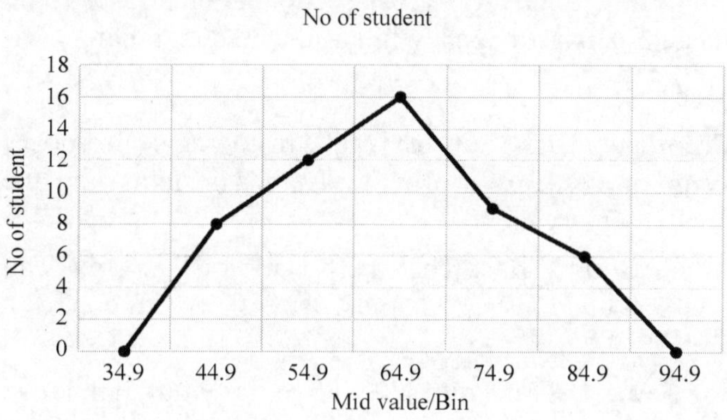

3.10.5.6 Cumulative Frequency Polygon or Ogive

A cumulative frequency polygon, also known as an ogive, is a statistical graph that represents the cumulative frequencies of a dataset. Unlike a standard frequency polygon that depicts frequencies of individual classes, an ogive shows the running total of frequencies up to each class boundary. This makes it an invaluable tool for understanding the distribution of data and for determining quantiles, such as medians or percentiles.

When to Use an Ogive

i) **To Determine Quantiles:** They allow for easy determination of medians, quartiles, and percentiles by visually inspecting where certain percentages of the data fall.

ii) **To Analyze the Distribution:** Ogives help in understanding the overall distribution of the data, including where the bulk of the data lies and how it spreads out.

iii) **Comparative Analysis:** When comparing the cumulative frequency distributions of two or more datasets, ogives provides a clear visual comparison.

Advantages of Ogives

i) **Clear Visualization:** Ogives offer a clear and straightforward way to visualize how data accumulates over a range, which can be especially helpful in identifying distribution patterns and trends.

ii) **Easy to Determine Quantiles:** They make it simple to visually estimate the median, quartiles, and specific percentiles, aiding in the understanding of data distribution. For example: the cross-sectional point of less than and more than the cumulative frequency line is the median point.

Disadvantages of Ogives

i) **Less Detail on Frequency Distribution:** Unlike histograms or standard frequency polygons, ogives do not provide detailed insights into the frequency of individual intervals.

ii) **Dependent on Class Boundaries:** The interpretation of an ogive can vary with how class boundaries are defined, potentially affecting the accuracy of quantile estimations.

iii) **Less Intuitive for Some Users:** For those unfamiliar with cumulative frequency graphs, ogives may be less intuitive to interpret compared to other graphical representations like histograms.

Example: Construct a cumulative frequency polygon or ogive based on the following data:

Marks	0-20	20-40	40-60	60-80	80-100
No of student	5	7	20	36	18

Solution

Marks	No of student	Lower boundary	Less than CF	More than CF
0-20	5	0	0	86
20-40	7	20	5	81
40-60	20	40	12	74
60-80	36	60	32	54
80-100	18	80	68	18
		100	86	0

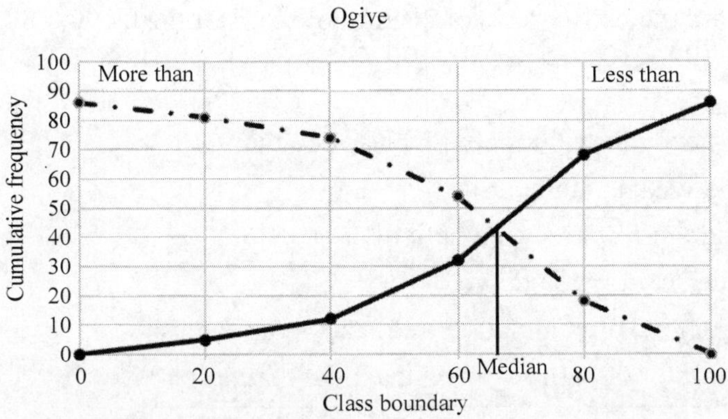

3.11 Chapter Summary

We began by unravelling the distinction between primary and secondary data collection and understanding their respective advantages and disadvantages. Primary data emerged as the cornerstone of empirical research, offering authenticity and relevance, while secondary data is presented a resourceful means of tapping into existing information. The steps involved in constructing a frequency distribution illuminates the significance of data organization and summarization, paving the way for more meaningful analysis.

The chapter then transitioned to the realm of data presentation, where we explored the art of translating raw data into visual representations. Pie charts, histograms, bar and column charts, and line charts have been emerged as versatile tools that enhance data comprehension and encourage informed decision-making. Each technique caters to distinct types of data and analytical objectives, allowing professionals to present their findings in a manner that resonates with their audience.

3.12 Self-Assessment

1. What are the advantages and disadvantages of the primary data collection method?
2. What are the advantages of secondary data collection and give some sources of secondary data?
3. Write short notes about: (i) frequency distribution, (ii) cumulative frequency distribution, and (iii) relative frequency distribution.
4. What are the steps involved in the preparation of a frequency distribution?
5. Create a frequency distribution table with the following data:

   7, 10, 12, 7, 5, 12, 15, 8, 9, 14, 17, 10, 11, 13, 14, 15, 17, 20, 18, 8, 9, 10

6. The following are the daily wages of 20 employees: 180, 190, 190, 180, 200, 220, 170, 195, 210, 201, 230, 220, 222, 185, 195, 201, 222, 227, 208, 202.
   i) Create a frequency distribution with class boundary.
   ii) Calculate relative and cumulative frequency.
   iii) Draw a histogram from the calculations.
   iv) Draw a frequency polygon curve.
   v) Construct an Ogive or cumulative frequency polygon.
7. Draw a line graph for the following data and label it properly.

Age	18	19	20	21	22	23
No of student	5	18	20	27	22	6

8. Draw a pie chart for the following data:

Investment plan	**Agriculture**	**Education**	**Industry**	**Communication**
Amount (Rs. Crores)	8000	6000	3000	1000

9. A car company manufactures 3 different models of car. The sale for the last six years is presented below. Draw a grouped bar chart for the following data:

Month	**Sales**		
	Model 1	**Model 2**	**Model 3**
2018	20	14	15
2019	25	15	15
2020	26	16	16
2021	28	14	12
2022	18	10	16
2023	9	6	19

10. During 2021-2024, the number of students in a university is as follows. Draw a stacked bar chart for the following data.

Year	**Commerce**	**Science**	**Arts**
2021	120	250	380
2022	140	280	300

Unit-IV

Measures of Central Tendency and Partition Values

4.1 Learning Objectives

By the end of this chapter, you will be able to:

i) Understand the concepts of Central Tendency and Partition Values;

ii) Identify appropriate measures for different data sets;

iii) Distinguish between different types of means;

iv) Apply Partition Values in data analysis.

4.2 Introduction

Nowadays, the ability to proficiently analyze and interpret data sets is indispensable for strategic decision-making. This chapter, 'Measures of Central Tendency and Partition Values,' extends beyond the basics of mean, median, and mode to explore the deeper layers of data analysis through quartiles, deciles, and percentiles. These measures offer a more comprehensive understanding of data distribution, crucial for identifying trends and making precise business predictions. We will unravel how these statistical tools can be effectively employed to dissect and interpret complex data sets, providing business professionals with an enhanced toolkit for informed analysis and decision-making in various business scenarios.

4.3 Measures of Central Tendency

Measures of Central Tendency are statistical tools used to identify the single value that best represents the central point within a dataset. This value is a summary of the entire data set, providing a snapshot of the typical or average value around which other values tend to cluster. These measures are fundamental in business analysis as they offer a simplified overview of complex data, enabling business professionals to make informed decisions quickly.

The importance of measures of central tendency in business cannot be overstated. They help in setting standards, understanding typical values in sales, revenues, customer behavior, and market trends, and are crucial in comparing different sets of data over time. By understanding the central point of data sets, businesses can identify norms and anomalies, aiding in strategic planning and operational adjustments.

There are three main types of measures of central tendency: **Mean**, **Median**, and **Mode**. Each has its unique method of calculation and application, and understanding when and how to use each measure is key to accurate analysis.

4.3.1 Mean

The mean is one of the most commonly used measures of central tendency in statistical analysis. It represents the average of a dataset, providing a single value that summarizes the entire set of numbers. The mean is calculated by adding all the numerical values in the dataset and dividing this total by the number of values.

Mean is primarily three types: **(a)** Arithmetic Mean; **(b)** Geometric Mean; and **(c)** Harmonic Mean.

4.3.1.1 Arithmetic Mean

The arithmetic mean (A.M.) is the most common type of mean used in statistical analysis. It is defined as the sum of all the values in a dataset divided by the number of values in that dataset. This measure provides a central value for the data, often referred to simply as the 'average'.

When to Use A.M.

i) **Interval and Ratio Data:** The arithmetic mean is the best suited for data measured on interval or ratio scales, where the intervals between values are consistent and meaningful. This includes data like temperatures (in Celsius or Fahrenheit), annual incomes, or heights.

ii) **Symmetric Distributions:** It is most effective when the data distribution is symmetric i.e. not skewed significantly to the left or right. In such distributions, the mean, median, and mode are approximately equal, and the mean provides a true center.

iii) **Homogeneous Data:** The arithmetic mean is particularly useful when the data points are relatively similar and there are no extreme values (outliers). In such cases, it accurately reflects the central tendency of the data.

Advantages of Arithmetic Mean

i) **Simplicity and Familiarity:** It is straightforward to calculate and widely understood, making it a popular choice for general data analysis.

ii) **Basis for Other Statistical Measures:** Many statistical methods and tests rely on the mean as a foundational element.

iii) **Efficiency:** The arithmetic mean can be used to summarize large data sets with a single value, making it efficient for quick overviews and comparisons.

Disadvantages of Arithmetic Mean

i) **Sensitivity to Outliers:** The arithmetic mean can be significantly skewed by outliers, which may lead to misleading results. For example, a single very high income can raise the average income in a way that doesn't accurately reflect the majority of incomes in a dataset.

ii) **Not Always Representative:** In skewed distributions, the arithmetic mean might not be the best measure of central tendency, as it might not lie at the center of the data.

Calculation Procedure

Simple Arithmetic Mean

When a variable x takes on n distinct values , then the arithmetic mean of x is defined as

$$\overline{x} = \frac{x_1 + x_2 + \cdots + x_n}{n} = \frac{\sum_{i=1}^{n} xi}{n}$$

Example: Compute A.M. of a set of numbers 7, 10, 6, 8, 12.

$$\overline{x} = \frac{\sum x}{n} = \frac{7+10+6+8+12}{5} = 8.6$$

Weighted Arithmetic Mean

When a variable x takes on n distinct values , each with corresponding frequencies, the arithmetic mean (A.M.) of x is determined by the formula:

$$\overline{x} = \frac{f_1x_1 + f_2x_2 + \cdots + f_nx_n}{f_1 + f_2 + \cdots + f_n} = \frac{\sum_{i=1}^{n} f_i x_i}{N}$$

Where $N = \sum_{i=1}^{n} fi$

Example: Compute A.M. based on the data as follows:

Value	1	2	3	4	5	6	7	8
Frequency	5	8	7	11	17	20	14	2

Solution

Value (x)	Frequency (f)	
1	5	5
2	8	16
3	7	21
4	11	44
5	17	85
6	20	120
7	14	98
8	2	16
Total	84	405

$$\overline{x} = \frac{\sum f_x}{N} = \frac{405}{84} = 4.82$$

Arithmetic Mean of Grouped Data

For the calculation of the arithmetic mean (A.M.) of grouped data, it is assumed that all data points are concentrated at the midpoint of their respective class intervals. Hence, if represent these midpoints for the class intervals with corresponding frequencies , the A.M. is defined as

$$\overline{x} = \frac{\sum f_i x_i}{N}$$

Where $N = \Sigma fi$

Example: Compute A.M. from the following data:

Class	0-10	10-20	20-30	30-40	40-50
Frequency	5	8	14	9	6

Solution

Calculation of Arithmetic Mean

Class	Frequency (f)	Mid-value (x)	
0-10	5	5	25
10-20	8	15	120
20-30	14	25	350
30-40	9	35	315
40-50	6	45	270
Total	42		1080

Arithmetic Mean:

Step Division Method of Calculating A.M. (Equal Width)

When dealing with large numbers for ease of computation, the variable x is transformed into d using the formula $di = \frac{xi - A}{I}$; where $i = 1,2, ..., n$.

Subsequently, the A.M. formula is as follows: $\bar{x} = A + \frac{\sum fd}{N} \times I$; where A is known as "change in origin (or assumed mean) and is known as "change in scale" or class width.

It's worth noting that A is typically selected as the observation situated at or near the midpoint of the distribution, while I is usually chosen as either the **class width** for intervals of **equal width** or the **Highest Common Factor** (H.C.F.) of the class widths for intervals of **unequal width**. The Highest Common Factor (HCF), commonly referred to as the Greatest Common Divisor (GCD), refers to the greatest common divisor of two or more integers, which is the biggest positive integer that can divide them without any remainder.

Example 1: Compute A.M. from the following data:

Height (cm)	125-129	130-134	135-139	140-144	145-149	150-154	155-159	160-164
Frequency	3	8	9	12	14	9	6	2

Solution

Class	Frequency (f)	Mid-value (x)	[A=147, I=5]	
125-129	3	127	-4	-12
130-134	8	132	-3	-24
135-139	9	137	-2	-18
140-144	12	142	-1	-12
145-149	14	147	0	0
150-154	9	152	1	9
155-159	6	157	2	12
160-164	2	162	3	6
Total	63			-39

Step Division Method of Calculating A.M. (Unequal Width)

Example 2: Compute arithmetic mean from the following data:

Class	0-4	5-14	15-19	20-34	35-39
Frequency	3	10	8	16	7

Solution

Let, [5 is the H.C.F of all the class widths]

Class	Frequency (f)	Mid-value (x)	[A=17, I=5] $d_i = \frac{x_i - A}{I}$	*fd*
0-4	3	2	-3	-9
5-14	10	9.5	-1.5	-15
15-19	8	17	0	0
20-34	16	27	2	32
35-39	7	37	4	28
Total	44			36

$$\overline{x} = \frac{n_1 x_1 + n_2 \overline{x}_2}{n_1 + n_2}$$

Arithmetic Mean of the Composite Group

The technique is particularly useful when dealing with a composite group, which consists of different subgroups, each with their own data set. The calculation procedure for the arithmetic mean of a composite group involves taking into account the means and sizes of the individual subgroups.

Let, two subgroups have a number of observations and and A.M.s and , then the composite A.M. of these groups is given by

$$\overline{x} = \frac{n_1 \overline{x}_1 + n_2 x_2}{n_1 + n_2}$$

For three groups, the corresponding formula is

$$\overline{x} = \frac{n_1 \overline{x}_1 + n_2 \overline{x}_2 + n_3 \overline{x}_3}{n_1 + n_2 + n_3}$$

Example: Suppose we have two subgroups A and B. Group A has 10 members with a mean score of 70, and Group B has 15 members with a mean score of 80.

Solution

Here, $x_1 = 70, n_1 = 10, x_2 = 80, n_2 = 15$,

$$\overline{x} = \frac{n_1 \overline{x}_1 + n_2 \overline{x}_2}{n_1 + n_2} = \frac{(70 \times 10) + (80 \times 15)}{10 + 15} = \frac{1900}{25} = 76$$

Important Properties of A.M.

i) The total of a set of observations is equal to the product of their number and the A.M.

a) $\sum x_i = n\overline{x}$; b) $\sum f_i x_i = N\overline{x}$

Assuming each observation is distinct, the initial equation (a) indicates that the simple sum is equal to the product (multiplication) of their number and the simple A.M. The subsequent equation (b) shows that the weighted sum is equal to the product of the sum of weights and the weighted A.M.

ii) The sum of the deviations of a set of observations from their A.M. is always zero.

If a variable x has n distinct values with corresponding frequencies, then their A.M. is expressed as:

$$\overline{x} = \frac{\sum f_i x_i}{N} \quad \dots (1)$$

Where

The total deviation of these observations from is calculated as:

$$D = f_1(x_i - \overline{x}) + f_2(x_2 - x) + \cdots + f_n(x_n - \overline{x})$$

$$= \sum_{i=1}^{n} f_i(x_i - \overline{x}) = \sum f_i x_i - \sum f_i \overline{x} = N\overline{x} - \overline{x}\sum f_i \text{ [Since from (1) } \sum f_i x_i = N\overline{x}]$$

$$= N\overline{x} - N\overline{x} = 0 [f_i = N]$$

[Since from (1)]

[Since]

iii) The sum of the squares of deviations of a set of observations has the smallest value when deviations are taken from their A.M.

If a variable x has n distinct values with corresponding frequencies, then their A.M. is expressed as:

$$x = \frac{\sum f_i x_i}{N}; \text{where } N = \sum f_i$$

$$D = \sum_{i=1}^{n} f_i(x_i - A)^2$$

Now, consider calculating the deviation of all observations from a chosen arbitrary value, A. The sum of the squares of these deviations from A is represented as:

Our aim is to demonstrate that D attains its minimum value when. To prove this, we proceed as follows:

$$D=\sum f_i(x_i-A)^2=\sum f_i[(x_i-\overline{x})+(\overline{x}-A)]^2$$
$$=\sum f_i[(x_i-\overline{x})^2=[(\overline{x}-A)+^2(x_i-\overline{x})(\overline{x}-A)]$$
$$=\sum f_i(x_i-\overline{x})^2+\sum f_i(\overline{x}-A)^2+\sum f_i^2(x_i-\overline{x})(\overline{x}-A)$$
$$=\sum f_i(x_i-\overline{x})^2+(\overline{x}-A)^2.\sum f_i+2(\overline{x}-A).\sum f_i(x_i-\overline{x})$$
$$=\sum f_i(x_i-\overline{x})^2+(\overline{x}A)^2.N+0[\text{Since}\sum f_i(x_i-\overline{x})=0]$$

Thus, we find that D is equal to the sum of two non-negative terms of which the first one has a fixed value, therefore, D will be minimum. When the second term is minimum, i.e., zero. This is possible when $A=\overline{x}$

Therefore, we observe that D is the sum of two non-negative terms, where the first term has a constant value. Hence, D reaches its minimum when the second term is at its smallest, which is zero. This occurs when $A=\overline{x}$

Note: It is important to note that when , , i.e.

$$A\neq\overline{x},D>\sum f_i(x_i-\overline{x})^2,\text{ie.}\sum f_i(x_i-A)^2>\sum f_i(x_i-\overline{x})^2$$

iv) If , where A and I are two constants, prove that.

Let each represent the ith value of a variable x, having a frequency. *fi* (*i* = 1,2 ... , *n*) Consequently, the Arithmetic Mean (A.M.) of x is determined as follows

$$\overline{x}=\frac{\Sigma f_i x_i}{N};\text{Where}\,N=\Sigma f_i$$

Now, $\overline{x}=\frac{1}{N}\sum f_i x_i=\frac{1}{N}\sum f_i(A+d_iI)=\frac{1}{N}\sum f_iA+\frac{1}{N}\sum f_id_iI$

$$=A.\frac{1}{N}\sum f_i+I.\frac{1}{N}\sum f_id_i=A+I\overline{d}[\text{Since}\,N=\sum f_i\text{ and }\overline{d}=\frac{1}{N}\sum f_id_i]$$

$$=A+\overline{d}I$$

4.3.1.2 Geometric Mean

The geometric mean (G.M.) is a type of average where we multiply the values together and then take a root of the product. Specifically, for a set of numbers, the geometric mean is the nth root (where n is the count of numbers) of the product of the numbers.

When to Use the G.M.

The geometric mean is preferable for certain types of data, especially where the values are multiplicative in nature or involve rates of change. Here are some specific cases where using the geometric mean is more appropriate:

i) **Multiplicative Growth Rates:** It's ideal for averaging ratios, percentages, or growth rates. This includes financial data like investment returns, interest rates, or population growth rates. The geometric mean accurately reflects the compounding effect that occurs in these types of data.

ii) **Data Spanning Multiple Orders of Magnitude:** When data spans a wide range (e.g., 1, 100, 10,000), the geometric mean provides a more representative average because it reduces the impact of very large values. This is particularly useful in fields like biology, where data can vary significantly in scale.

iii) **Proportional or Relative Changes:** It's suitable for data that represents proportional or relative changes, as in the case of growth rates or normalized data.

iv) **Product of Variables:** In situations where the variables are multiplicatively related (such as length, area, and volume in geometry), the geometric mean gives a meaningful central value.

v) **Data that should be Log-Normally Distributed:** If the data is log-normally distributed (i.e., the logarithm of the data follows a normal distribution), the geometric mean is a natural choice for determining the central tendency.

vi) **Portfolio Performance:** In finance, to measure the performance of investment portfolios over time, especially when dealing with compound interest and reinvestments.

vii) **Harmonic Mean as a Special Case:** In certain scenarios, like calculating average speeds or rates, the geometric mean can be a more appropriate choice than the arithmetic or harmonic means.

Advantages of G.M.

i) **Reduces the Skewness Effect:** It minimizes the impact of extreme outliers. This is particularly beneficial for skewed distributions, as it provides a more representative average.

ii) **Balanced Representation of Data:** The geometric mean ensures that each data point contributes proportionally to the final average, preventing larger numbers from disproportionately influencing the result.

iii) **Handles Multiplicative Processes Well:** It accurately reflects situations where numbers are multiplied together and their relationships are multiplicative, such as in rates of growth.

iv) **Compounding Effects:** It correctly represents the nature of compounding effects, which is crucial in financial calculations like portfolio returns or interest rates.

v) **Meaningful in Ratio Analysis:** For averaging ratios or indices, the geometric mean provides a more relevant measure than the arithmetic mean.

Disadvantages of G.M.

i) **Inapplicability to Negative or Zero Values:** It cannot be used with datasets that include negative or zero values, limiting its applicability.

ii) **Computational Complexity:** Calculating the geometric mean is more mathematically complex, requiring exponentiation and nth roots, which may not be straightforward without computational tools.

iii) **Sensitivity to Small Values:** The geometric mean is more sensitive to small values, which can significantly lower the average in a way that might not accurately reflect the data distribution.

iv) **Difficulty in Further Analysis:** Conducting additional statistical analyses using the geometric mean (like variance or standard deviation) is more complex compared to the arithmetic mean.

v) **Less Intuitive Understanding:** For those unfamiliar with statistical concepts, the geometric mean can be less intuitive and harder to understand than the arithmetic mean.

vi) **Not a Measure of Central Tendency for All Distributions**: It is not always the most appropriate measure of central tendency, especially for data that does not involve growth rates, ratios, or proportions.

Calculation Procedure

Simple Geometric Mean (G.M.)

G.M. of a set of n numbers x_1, x_2, ..., x_n is defined as the nth root of their product.

Symbolically,

$$\text{G.M.} = \sqrt[n]{x_1.x_2....x_n} = (x_1.x_2.\,...x_n)^{\frac{1}{n}}$$

However, this formula cannot be applied directly for practical calculations. Taking logarithms of both sides we have

$$\log g = \frac{1}{n}(\log x_1 + \log x_2 + \cdots + \log x_n) = \frac{1}{n}\sum \log x$$

Example: Compute the G.M. of 5, 10, 18.

Solution

$$\log g = \frac{1}{n}(\log 5 + \log 10 + \log 18) = \frac{1}{3}(2.954) = 0.9846g$$

$$g = \text{antilog}(0.9846) = 9.6516$$

Weighted Geometric Mean (G.M)

Let $x_1, x_2, ..., x_n$ be n different values assumed by a variable x with respective frequencies. Then the G.M. of x is defined as

$$G = \sqrt[n]{x_1^{f_1} x_2^{f2} \cdots x_n^{fn}}$$

Or,

$$G = (x_1^{f_2} x_2^{f_2} \cdots x_n^{f_n})\frac{1}{N}, \text{where } N = \sum f_i$$

Practically, both side log is required to calculate the G.M. value. After taking the log, the formula can be expressed as:

$$\log G = \frac{1}{N}[f1(\log x_1) + f_2 (\log x_2) + \cdots + f_n (\log x_2)]$$

$$= \frac{1}{N}\sum f_i (\log x_i)$$

Example: Compute G.M. of 120, 118, 98, 109, 122, 200 having weights 4, 1, 3, 2, 5, 6.

Solution

x	f	$\log x$	$f(\log x)$
120	4	2.0792	8.3168
118	1	2.0718	2.0718
98	3	1.9912	5.9736
109	2	2.0374	4.0748
122	5	2.0864	10.432
200	6	2.3010	13.806
Total	21		44.675

$$\log G = \frac{1}{N}\sum fi(\log xi) = \frac{44.675}{21} = 2.1274$$

$$G = \text{anti}\log(2.1274) = 134.09$$

[Note: antilog (2.1274) = $10^{2.1274}$ = 134.09]

Geometric Mean for Composite Groups

Let there be two groups containing n_1 and n_2 observations and having respective G.M. G_1 and G_2. Then, the G.M. of the composite group is given by

$$G=(G_1^{n_1}.G_2^{n_2})\frac{1}{n_1+n_2}$$

Similarly, for three groups the corresponding formula is

$$G=(G_1^{n_1}.G_2^{n_2}.G_3^{n_3})\frac{1}{n_1+n_2+n_3}$$

Example: The G.M. of 5 numbers is 22 and the geometric mean of 8 others is 26. Compute G.M. for all 13 numbers together.

Solution

Let $n_1 = 5$, $G_1 = 22$, $n_2 = 8$, $G_2 = 26$. Therefore, the composite G.M. is given by

$$G=(G_1^{n1}.G_2^{n2})^{\frac{1}{n_1+n_2}}=(22^5.26^8)^{\frac{1}{13}}$$

$$\log G=\frac{1}{13}\log(22^5.26^8)=\frac{1}{13}(5\log 22+8\log 26)$$

$$=\frac{1}{13}(5\times1.3424+8\times1.4149)=\frac{1}{13}(6.712+11.3192)$$

$$=\frac{1}{13}18.0312=1.3870$$

$$G=\text{anti}\log(1.3870)=24.378$$

Properties of G.M.

i) The logarithm of the G.M. of a set of numbers is equal to the A.M. of the logarithms of those numbers.

The G.M. of a set of numbers with respective frequencies is defined as

$$G=(x_1^{f1}.x_2^{f2}.\cdots x_n^{fn})^{\frac{1}{N}},\text{ where } N=\sum \text{ft}$$

Or

$$\log G = \frac{1}{N}\log(x_1^{f1}.x_2^{f2}\cdots x_n^{fn})$$

$$= \frac{1}{N}(\log x_1^{f_1} + \log x_2^{f_2} + \cdots + \log x_n^{f_n})$$

$$= \frac{1}{N}[f_1(\log x_1) + f_2(\log x_2) + \cdots + f_n(\log x_n)]$$

$$= \frac{1}{N}\sum_{i=1}^{n} f_i(\log x_i)$$

Hence the result follows.

ii) If be the geometric means of several groups having observations respectively then G.M. (G) of the composite group is given by their weighted geometric mean.

$$G = \sqrt[n]{G_1^{n_1} G_2^{n_2} \cdots}$$

$$\log G = \frac{1}{N}\sum n_i(\log G_i)$$

Where, $N = n_1 + n_2 + \ldots$

iii) The product of a group of n observations is equal to the n-th power of their G.M.

G.M. of a set of n numbers $x_1, x_2, \ldots, x_n$ is defined as the nth root of their product. Symbolically,

4.3.1.3 Harmonic Mean

The harmonic mean (H.M) is defined as the reciprocal of the arithmetic mean of the reciprocals of a set of numbers. Mathematically, it is expressed as:

Where H is the harmonic mean, n is the number of observations in the dataset, and are the individual numbers in the dataset.

When to Use H.M.

The harmonic mean is particularly useful in specific scenarios, including:

i) **Averaging Ratios or Rates:** When you are dealing with rates, like speed, the harmonic mean provides a more accurate average than the arithmetic mean. This is because it accounts for the weighted contribution of each rate to the overall average.

ii) **Investment and Finance:** In finance, the harmonic mean can be used to calculate average price multiples, like the price-earnings ratio. It gives a more representative average when comparing the ratios of different companies.

iii) **Harmonizing Units:** When units are in the denominator, like in the case of speed (distance/time), the harmonic mean is appropriate. It harmonically averages the time component, which is more accurate for certain analyses.

Advantages of H.M.

i) **Accuracy for Rates:** Provides a more accurate average when dealing with rates or ratios, particularly when the data involves quantities in the denominator (like speed).

ii) **Mitigates Impact of Large Outliers:** Less influenced by large outlier values compared to the arithmetic mean, which can be useful in certain datasets to avoid skewed results. This is because it gives greater weight to smaller values, making it more robust in certain data sets with extreme values.

iii) **Robustness in Specific Contexts:** Offers a more representative average for specific types of data sets where arithmetic or geometric means are not suitable.

Disadvantages of H.M.

i) **Inapplicable with Zero Values:** This cannot be used if the dataset contains zero values, as it will lead to undefined or infinite results.

ii) **Limited Applicability:** Its usefulness is confined to specific types of data (like rates and ratios), making it less versatile compared to other means.

iii) **Complexity in Calculation:** Computing the harmonic mean is generally more complex and time-consuming than calculating the arithmetic mean, especially for large datasets.

iv) **Less Intuitive Understanding:** The concept and interpretation of the harmonic mean can be less intuitive for those not familiar with statistical methods.

Calculation Procedure

Simple Harmonic Mean (unweighted data)

Example: Compute the H.M. of the numbers 6, 8, 10, 4, 2

Solution

x	$\frac{1}{x}$
6	0.1667
8	0.125
10	0.1
4	0.25
2	0.5
Total	1.1417

$$H = \frac{n}{\sum \frac{1}{x}} = \frac{5}{1.1417} = 4.38$$

Weighted Harmonic Mean

$$H = \frac{f_1 + f_2 + \cdots + f_n}{\frac{f_1}{x_1} + \frac{f_2}{x_2} + \cdots + \frac{f_n}{x_n}} = \frac{N}{\sum_{i=1}^{n} \frac{f_i}{x_i}}$$

Where $N = \Sigma fi$

Example: Compute H.M. from the following data:

Class	10-14	15-19	20-24	25-29	30-34
f	6	8	10	4	2

Solution

Class	f	Mid value (x)	$\frac{f}{x}$
10-14	6	12	0.5
15-19	8	17	0.4706
20-24	10	22	0.4545
25-29	4	27	0.1481
30-34	2	32	0.0625
Total	30		1.6357

$$H = \frac{N}{\sum \frac{f}{x}} = \frac{30}{1.6357} = 18.34$$

4.3.1.4 Relation among A.M., G.M., and H.M.

For any two positive numbers, the relation will be: A.M. ≥ G.M. ≥ H.M.

Proof: Let and be two positive numbers.

$$\text{Therefore, their A.M.} = \frac{x_1 + x_2}{2}; \text{G.M.} = \sqrt{x_1 x_2} \text{ and H.M.} = \frac{2}{\frac{1}{x_1} + \frac{1}{x_2}}$$

Now, we know, $\left(\sqrt{x1} - \sqrt{x_2}\right)^2 \geq 0$ [Since a square quantity is always non-negative].

$$\text{Or, } x_1 + x_2 - 2\sqrt{x_1 x_2} \geq 0 \qquad \text{Or, } x_1 + x_2 \geq \sqrt{x_1 x_2} \qquad \text{Or, } \frac{x_1 x_2}{2} \geq \sqrt{x_1 x_2}$$

\ A.M. ≥ G.M.

$$\text{Again, } \left(\frac{1}{\sqrt{x_1}} - \frac{1}{\sqrt{x_2}}\right)^2 \geq 0 \qquad \text{Or, } \frac{1}{x_1} + \frac{1}{x_2} \geq \frac{2}{\sqrt{x_1 x_2}} \qquad \text{Or, } \sqrt{x_1 x_2} \geq \frac{2}{\frac{1}{x_1} + \frac{1}{x_2}}$$

\ G.M ≥ H.M.

From the above two findings, we can say that A.M. ≥ G.M. ≥ H.M.

For a pair of positive numbers A. M. x H. M. = G. M.^{2}

Proof: Let and be two positive numbers.

$$\text{Therefore, A.M.} = \frac{x_1 + x_2}{2}; \text{G.M.} = \sqrt{x_1 x_2} \text{ and H.M.} = \frac{2}{\frac{1}{x_1} + \frac{1}{x_2}} = \frac{2x_1 x_2}{x_1 + x_2}$$

$$\text{Now, A.M.} \times \text{H.M.} = \frac{x_1 + x_2}{2} \times \frac{2x_1 x_2}{x_1 + x_2} = x_1 x_2 = \left(\sqrt{x_1 x_2}\right)^2 = (\text{G.M.})^2$$

4.3.2 Median

The median is the value separating the higher half from the lower half of a data set. It is a form of measure of central tendency, like the mean and mode.

When to Use the Median

i) **Skewed Distributions:** The median is often used instead of the mean when the data distribution is skewed (not symmetric). In a skewed distribution, the mean can be distorted by outliers (extremely high or low values), but the median remains unaffected by them.

ii) **Ordinal Data:** The median is suitable for ordinal data (data that can be ranked but not measured, like survey responses: poor, fair, good, very good, excellent).

iii) **Describing the Typical Value:** When you want to describe the typical value in a dataset, especially when the data includes outliers or non-numeric data (like ratings or scales), the median can be more representative than the mean.

iv) **Income and Wealth Data:** In economics, the median is often used to provide a more accurate representation of typical income or wealth, as these distributions are usually highly skewed.

v) **Real Estate and Market Analysis:** Median values are frequently used in real estate to report central tendencies in house prices, as they are less influenced by extremely high or low values.

Advantages of Median

i) **Resistant to Outliers:** The median is not affected by extreme values (outliers). This makes it a more reliable measure of central tendency when the dataset contains outliers or is skewed.

ii) **Represents the Middle Value:** The median accurately divides the dataset into two equal halves. This can be useful in understanding the distribution of the data.

iii) **Useful for Ordinal Data:** It can be used with ordinal data (data which can be ranked or ordered) where mean or mode may not be appropriate.

iv) **Easy to Understand and Calculate:** For small datasets, it is straightforward to find the median by simply ordering the data and finding the middle value.

v) **Non-parametric Measure:** Since the median only depends on the order of values, not their exact magnitude, it is suitable for non-parametric statistics.

Disadvantages of the Median

i) **Not as Informative for Small Datasets:** In small datasets, the median might not provide a clear picture of the data distribution.

ii) **Affected by Sample Size:** An even number of data points requires averaging the two middle numbers, which can be misleading in certain cases.

iii) **Does Not Use All Data Points:** Unlike the mean, the median does not consider the value of each data point, which might lead to loss of information, especially in datasets with varying ranges.

iv) **Not Suitable for Further Statistical Analysis:** The median is not as amenable to algebraic manipulation as the mean, limiting its use in further statistical calculations.

v) **Inefficiency in Large Datasets:** Calculating the median in large datasets can be more cumbersome as it requires the data to be sorted first.

Calculation Procedure

Median for Ungrouped Data

$$\text{Median} = \text{the value of the } \frac{N+1}{2} \text{th item.}$$

Where N= Total number of items.

Example: Find the median value of the following numbers: 15, 34, 45, 23, 44, 67, 64, 86

Solution

We first arrange the numbers in ascending order 15, 23, 34, 44, 45, 64, 67, 86

The number of observations (N) = 8

$$\text{Median} = \frac{N+1}{2}\text{th item} = \frac{8+1}{2} = 4.5\text{th item}$$

$$= \frac{[4\text{th item} + 5\text{th item}]}{2} = \frac{44+45}{2} = 44.5$$

Median for Grouped Data

$$\text{Median} = L + \frac{\frac{N}{2} - cf}{f} \times I$$

Where, L = Lower boundary of the median class,

cf = cumulative frequency (less than) preceding the median class,

f = frequency of the median class,

I = width of the median class,

N = Sum of frequency (Σf_i).

Example: Compute the median value for the following data:

Weight (kg)	31-35	36-40	41-45	46-50	51-55	56-60	61-65
No of Person	10	22	32	45	50	43	26

Solution

Class limit	Class boundaries	Frequency (f)	Cumulative frequency
31-35	30.5-35.5	10	10
36-40	35.5-40.5	22	32
41-45	40.5-45.5	32	64

46-50	45.5-50.5	45	109
51-55	**50.5-55.5**	**50**	**159**
56-60	55.5-60.5	43	202
61-65	60.5-65.5	26	228
		228 (N)	

Median class = The value of N/2th item

= The value of 114 (228/2) th item which lies in the class (50.5-55.5)

$$\text{Median} = L + \frac{\frac{N}{2} - cf}{f} x1 = 50.5 + \frac{114 - 109}{50} x\,5$$

$= 50.5 + 1.3 = 51.8\,kg$

Note: The median class is determined by the cumulative frequency (less than). The median value calculation process will be the same for the unequal width-based group data.

4.3.3 Mode

The mode in statistics is the value that appears most frequently in a data set. It represents the most common or popular item in the collection of data. Unlike the mean and median, the mode can be used with any type of data - numerical, categorical, ordinal, or nominal. It's particularly useful in understanding the distribution of non-numeric data.

When to Use Mode

i) **Categorical Data:** The mode is especially useful for categorical data where numerical means and medians are meaningless. For example, in a survey asking for people's favorite color, the mode identifies the most popular color choice.

ii) **Nominal Data:** With nominal data (data that is categorized without a natural order, like brands of cereal), the mode can help to identify the most common category.

iii) **Bimodal or Multimodal Distributions:** In distributions where two or more values appear with the same highest frequency, the dataset is described as bimodal or multimodal, respectively. The mode can help to identify these peaks in data.

iv) **Data with Repetitions:** When data values are repeated frequently, the mode provides insight into the most common value. This can be more informative than the mean or median in some cases.

v) **Non-Numeric Data:** For data that can't be quantitatively measured (like types of fruits in a basket), the mode is the only measure of central tendency that can be used.

vi) **Skewed Distributions:** In skewed distributions, the mode can be a better measure of central tendency than the mean, as it is not affected by extreme values.

Advantages of the Mode

i) **Applicability to Any Data Type:** The mode can be used with any level of data - nominal, ordinal, interval, or ratio. This makes it versatile, especially for non-numeric and categorical data where the mean and median are not applicable.

ii) **Identifies Most Common Value:** The mode reflects the most common or popular item in a dataset, providing clear insights into the preferences or trends within the data.

iii) **Unaffected by Extreme Values:** Unlike the mean, the mode is not influenced by outliers or extremely skewed data, making it a reliable measure in these situations.

iv) **Simple to Understand and Compute:** The concept of the mode is straightforward for most people to understand. It does not require any complex calculations, making it accessible without advanced statistical knowledge.

v) **Useful in Market Research and Economics:** In fields like market research, fashion, and economics, the mode can be particularly useful for identifying the most popular items, trends, or preferences.

Disadvantages of the Mode

i) **Not Always Representative:** The mode may not be representative of all the data points, especially in a dataset with a large range of values.

ii) **Possible to Have No Mode or Multiple Modes:** Some datasets may not have a mode (no repeats), while others can be bimodal or multimodal (having two or more modes), which can make interpretation more complex.

iii) **Limited Use in Further Statistical Analysis:** The mode is not as useful in advanced statistical analyses and mathematical operations as the mean or median.

iv) **Can Be Misleading in Small Samples:** In small datasets, the mode can be easily skewed by a slight repetition of certain values, leading to misleading conclusions.

v) **Does Not Reflect the Distribution of All Data Points:** The mode only considers the most frequent value and ignores the distribution of the rest of the data, which can lead to an incomplete understanding of the dataset.

Calculation Procedure

From a simple frequency distribution, the mode can be determined by inspection only. It is that value of the variable which corresponds to the largest frequency.

Example: calculate the mode of the following series:

a) 4, 5, 7, 6, 5, 8, 5, 9, 3, 1, 5

b) 8, 10, 12, 8, 3, 5, 10, 6, 8, 10

c) 1, 4, 5, 7, 8, 10, 12, 24, 32

Solution

a) It is observe that 5 occurs the maximum number of times (4 times). So, 5 is the mode.

b) In this case, we observe that there are two modes 8 and 10 as both of them occur a maximum number of times (3 times).

c) In this case, all items occur an equal number of times. So, there no mode value exists.

Mode for Group Data

Mode can be obtained both graphically and statistically.

Graphical Method: In graphical method, the mode can be obtained from the histogram. The steps are as follows:

Step 1: Construct a histogram for the provided data series.

Step 2: Identify the highest rectangle in the histogram; this represents the modal class.

Step 3: Connect the top left corner of the modal class's rectangle to the top left corner of the following class's rectangle with a straight line.

Step 4: Similarly, draw a line from the top right corner of the modal class's rectangle to the top right corner of the previous class's rectangle.

Step 5: At the intersection of the two lines drawn in steps 3 and 4, drop a perpendicular line to the X-axis.

Step 6: The point where this perpendicular line intersects the X-axis indicates the mode of the data series.

Example: Find the mode from the following data:

Class	0-10	10-20	20-30	30-40	40-50	50-60	60-70	70-80
Frequency	4	5	10	20	15	8	10	18

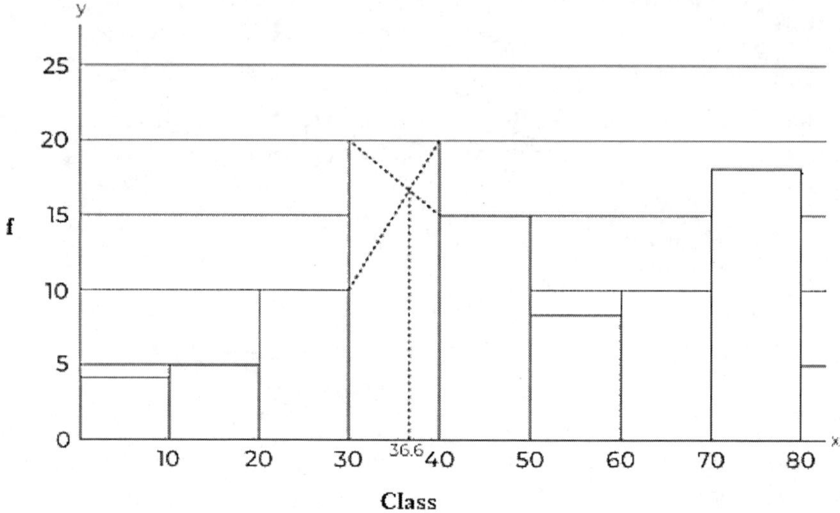

From the histogram, it is observed that the mode value is 36.6.

Note: if any dataset contains two or more highest frequencies then the graphical method will not be applicable.

Statistical method

The formula for calculating mode is as follows:

$$\text{Mode} = l_1 + \frac{d_1}{d_1 + d_2} \times c$$

Where,

l_1 = lower boundary of the modal class (i.e. the class containing the largest frequency);

d_1 = difference of the largest frequency and the frequency of the class just preceding the modal class ($d_1 = f_0 - f_{-1}$);

d_2 = difference between the largest frequency and the frequency of class just following the modal class ($d_1 = f_0 - f_1$);

c = common width of classes.

The formula of mode may also be written in the form as:

$$\text{Mode} = l_1 + \frac{f_0 - f_{-1}}{2f_0 - f_{-1} - f_1}$$

Here, f_0, f_{-1}, f_1, represent the frequency of the modal class, the class just preceding, and the class just following the modal class.

Example 1: Compute mode of the following data

Class	0-10	10-20	20-30	30-40	40-50	50-60	60-70	70-80
Frequency	4	5	10	20	15	8	10	18

Here, 30-40 is the modal class as it contains the highest frequency.

$l_1 = 30, f_0 = 20,\ f_{-1} = 10,\ f_1 = 15, c = 10$

$[d_1 = f_0 - f_{-1} = 20 - 10 = 10]; [d_2 = f_0 - f_1\ 20 - 15 = 5]$

$$\text{Mode} = l_1 + \frac{d_1}{d_1 + d_2} \times c$$

$$\text{Mode} = 30 + \frac{10}{10+5} \times 10 = 30 + 6.67 = 36.67$$

From the above graphical and statistical methods, we observe that the result is very similar.

Example 2: Compute mode value from the following data.

Marks	20-29	30-39	40-49	50-59	60-69	70-79	80-89
No of student	6	16	17	25	15	12	8

Solution

Class limit	**Class boundaries**	***f***	**Mid value (x)**	**Cum. f**
20-29	19.5-29.5	6	24.5	6
30-39	29.5-39.5	16	34.5	22
40-49	39.5-49.5	17	44.5	39
50-59	49.5-59.5	25	54.5	64
60-69	59.5-69.5	15	64.5	79
70-79	69.5-79.5	12	74.5	91
80-89	79.5-89.5	8	84.5	99
Total		99		

$$\text{Mode} = l_1 + \frac{d_1}{d_1 + d_2} \times c = 49.5 + \frac{(25-17)}{(25-17)+(25-15)} x10 = 49.5 + \frac{8}{18} \times 10 = 53.94$$

Mode for Unequal Width

The formula mentioned above can only be applied in situations where each class interval is of the same size and there is a single modal class present. If the class widths are unequal or there are multiple modal classes, then one must resort to using the empirical relationship among the mean, median, and mode to determine the mode. This relationship is expressed through a specific formula that is:

Mean – Mode =3 (Mean – Median)

Example: Calculate mode for the following data:

Class	5-14	15-19	20-29	30-39	40-44	45-49
f	4	6	10	15	8	3

Solution

Class	Class boundaries	f	x	fx	cf
5-14	4.5-14.5	4	9.5	38	4
15-19	14.5-19.5	6	17	102	10
20-29	19.5-29.5	10	24.5	245	20
30-39	29.5-39.5	15	34.5	517.5	35
40-44	39.5-44.5	8	42	336	43
45-49	44.5-49.5	3	47	141	46
		46		1379.5	

$$\text{Mean} = \frac{\sum fx}{N} = \frac{1379.5}{46} = 29.99$$

N/2=46/2=23 which lies in the class 29.5-39.5.

$$\text{Median} = L + \frac{\frac{L}{2} - cf}{f} \times I = 29.5 + \frac{23-20}{15} \times 10 = 29.5 + 2 = 31.5$$

Now, Mean – Mode = 3 (Mean – Median)

29.99 - Mode = 3(29.99 - 31.5)

- Mode = - 4.53 - 29.99

Mode = 34.52

4.4 Partition Values-Quartiles, Deciles, and Percentiles

Partition values in statistics, namely quartiles, deciles, and percentiles, play a pivotal role in providing a comprehensive understanding of data distribution. These measures divide a dataset into equal parts, offering a detailed perspective on its spread and variability.

The use of partition values in business statistics is invaluable. They aid in understanding the distribution of business-related metrics, such as customer income levels, sales figures, or production costs. By identifying where the majority of values lie within these partitions, businesses can tailor strategies, anticipate market trends, and make informed decisions.

4.4.1 Quartiles

Quartiles are statistical values that divide a variable into four equal parts, each representing 25% of the data. These quartiles are denoted as Q1, Q2, and Q3.

Q1 **(First Quartile):** This value marks the 25th percentile of the data. It means that 25% of the data points are less than or equal to Q1, and 75% are greater.

Q2 **(Second Quartile):** Commonly known as the **median**, Q2 represents the middle value of the dataset when arranged in ascending order. 50% of the data points are below Q2, and 50% are above.

Q3 **(Third Quartile):** This value signifies the 75th percentile of the data. 75% of the data points fall below Q3, and 25% are greater.

When to Use Quartiles

i) **Data Analysis and Visualization:** Quartiles are handy in summarizing the central tendency and dispersion of data. Box plots, for instance, often use quartiles to display the spread of the data.

ii) **Identifying Outliers:** Quartiles assist in identifying potential outliers in a dataset by comparing values that fall significantly below Q1 or above Q3.

iii) **Comparing Distributions:** They aid in comparing the distribution of different datasets, helping to understand similarities and differences in their spread and variability.

Calculation Procedure

For Ungrouped Data

Q_i = The value of $\frac{i(N+1)}{4}$ th item (i=1, 2, 3), where N = total number of observations.

Example: calculate all the quartiles based on the following data:

384, 390, 592, 418, 668, 534, 769, 754, 2200, 1500

Solution

The given values are arranged in increasing order of magnitude and ranked. Here N =10.

$$\text{Rank of } Q_1 = \frac{1}{4}(10+1) = 2.75; Q_2 = \frac{2}{4}(10+1) = 5.5; Q_a = \frac{3}{4}(10+1) = 8.25$$

Rank	Value
1	384
2	**390**
3	**418**
4	534
5	**592**
6	**668**
7	754
8	**769**
9	**1500**
10	2200

Using a simple interpolation method, we obtain

For Grouped Data

$$Q_i = L + \frac{\frac{iN}{4} - cf}{f} \times 1$$

Where, L = lower boundary of the quartile class,

f = frequency of the quartile class,

cf = cumulative frequency preceding the quartile class,

I = width of the quartile class.

Example: Compute all the quartiles from the following data:

Class	10-15	15-20	20-25	25-30	30-40	40-50	50-60	60-70
Frequency	4	12	17	23	11	9	7	3

Solution

Class boundaries	Frequency (f)	Cumulative frequency (cf)
10-15	4	4
15-20	12	16
20-25	17	33
25-30	23	56
30-40	11	67
40-50	9	76
50-60	7	83
60-70	3	86
	86	

Q_1 = The value of $\frac{iN}{4} = \frac{1}{4}$ th item = The value of 21.5 th item which lies in the class 20-25

$$Q1 = L + \frac{\frac{N}{4} - cf}{f} \times I = 20 + \frac{21.5 - 16}{17} \times 5 = 20 + 1.62 = 21.62$$

Q_2= The value of th item = The value of 43 th item which lies in the class 25-30

$$Q_2 = L + \frac{\frac{2N}{4} - cf}{f} \times 1 = 25 + \frac{43 - 33}{23} \times 5 = 25 + 2.17 = 27.17$$

Q_3= The value of th item = The value of 64.5 th item which lies in the class 30-40

$$Q_2 = L + \frac{\frac{3N}{4} - cf}{f} \times I = 30 + \frac{64.5 - 56}{11} \times 10 = 30 + 7.73 = 37.73$$

4.4.2 Deciles

Deciles divide a dataset into ten equal parts, where each decile represents 10% of the data. For instance, D_1 represents the 10th percentile, D_2 the 20th percentile, and so on until D_9, which represents the 90th percentile.

When to Use Deciles

i) **Data Analysis with More Detail:** Deciles provide a finer level of granularity than quartiles, making them useful when a more detailed understanding of data distribution is required.

ii) **Performance Analysis:** In various fields like finance or rankings, deciles help in assessing performance against segmented portions of the population. For instance, analyzing income distribution across deciles can reveal income disparities.

iii) **Market Segmentation:** Deciles can assist in market segmentation by dividing consumers into groups based on their spending habits, preferences, or demographics.

Calculation Procedure

The calculation of deciles is very similar to the calculation procedure of quartiles.

For Ungrouped Data

D_i= The value of $\frac{i(N+1)}{10}$ th item (i=1, 2, 3, …, 9) where N = total number of observations.

Example: compute the 2nd and 7th Decile for the following data:

384, 390, 592, 418, 668, 534, 769, 754, 2200, 1500

Solution

The given values are arranged in increasing order of magnitude and ranked. Here N = 10.

Rank of $D_2 = \frac{2}{10}(10+1) = 2.2; D_7 = \frac{7}{10}(10+1) = 7.7$

Rank	Value
1	384
2	**390**
3	**418**
4	534
5	592
6	668
7	**754**
8	**769**
9	1500
10	2200

Using a simple interpolation method, we obtain

$$\frac{2.2-2}{3-2} = \frac{D_2 - 390}{418-390}; D_2 = 5.6 + 390 = 395.6$$

$$\frac{7.7-7}{8-7} = \frac{D_7 - 754}{769-754}; D_7 = 10.5 + 754 = 764.5$$

For Grouped Data

$$D_i = L + \frac{\frac{iN}{10} - cf}{f} \times I$$

Where, = lower boundary of the decile class,

f = frequency of the decile class,

cf = cumulative frequency preceding the decile class,

I = width of the decile class.

Example: Compute the 6th and 9th deciles from the following data:

Class	10-15	15-20	20-25	25-30	30-40	40-50	50-60	60-70
Frequency	4	12	17	23	11	9	7	3

Solution

Class boundaries	Frequency (f)	Cumulative frequency (cf)
10-15	4	4
15-20	12	16
20-25	17	33
25-30	23	56
30-40	11	67
40-50	9	76
50-60	7	83
60-70	3	86
	86	

D_6 = The value of $\frac{6N}{10}=\frac{6}{10}\times 86$ th item = The value of 51.6 th item which lies in the class 25-30

$$D_6 = L + \frac{\frac{6N}{10}-cf}{f} \times I = 25 + \frac{51.6-33}{23} \times 5 = 25 + 4.04 = 29.04$$

D_9= The value of th item = The value of 77.4 th item which lies in the class 50-60

$$D_9 = L + \frac{\frac{9N}{10}-cf}{f} \times I = 50 + \frac{77.4-76}{7} \times 10 = 50 + 2 = 52$$

4.4.3 Percentiles

Percentiles divide a dataset into hundred equal parts. The nth percentile represents the value below which n% of the data falls. For instance, the 25th percentile (also known as the first quartile Q1) is the value below which 25% of the data falls.

When to Use Quartiles

i) **Comparing Individual Scores or Values:** Percentiles help to compare individual scores or values against a larger group. For instance, if you score in the 90th percentile on a test, it means your score is higher than 90% of the test-takers.

ii) **Assessing Growth and Development:** In healthcare, percentiles are used in growth charts to track a child's growth against standard measures. For example, a child in the 75th percentile for weight is heavier than 75% of children of the same age.

iii) **Benchmarking and Rankings:** In various fields like finance or demographics, percentiles help to understand where a particular data point stands in relation to the entire dataset. For instance, a company's revenue in the 95th percentile might indicate exceptional performance compared to its competitors.

iv) **Identifying Outliers:** Percentiles can assist in identifying outliers by highlighting values that fall significantly above or below certain percentile thresholds.

Calculation Procedure

The calculation of percentiles is very similar to the calculation procedure of quartiles.

For Ungrouped Data

P_i = The value of $\frac{i(N+1)}{100}$ th item (i=1, 2, 3, …, 99) where N = total number of observations.

Example: compute the 20th and 70th Percentile for the following data:

384, 390, 592, 418, 668, 534, 769, 754, 2200, 1500

Solution

The given values are arranged in increasing order of magnitude and ranked. Here n=10.

$$\text{Rank of } P_{20} = \frac{20}{100}(10+1) = 2.2; P_{70} = \frac{70}{100}(10+1) = 7.7$$

Rank	Value
1	384
2	**390**
3	**418**
4	534
5	592
6	668
7	**754**
8	**769**
9	1500
10	2200

Using a simple interpolation method, we obtain

$$\frac{2.2-2}{3-2} = \frac{P_{20}-390}{418-390}; P_{20} = 5.6 + 390 = 395.6$$

$$\frac{7.7-7}{8-7}=\frac{P_{70}-754}{769-754};P_{70}=10.5+754=764.5$$

For Grouped Data

$$P_i=L+\frac{\frac{iN}{100}-cf}{f}\times 1$$

Where, L = lower boundary of the percentile class,

f = frequency of the percentile class,

cf = cumulative frequency preceding the percentile class,

I = width of the percentile class.

Example: Compute the 60th and 90th percentile from the following data:

Class	10-15	15-20	20-25	25-30	30-40	40-50	50-60	60-70
Frequency	4	12	17	23	11	9	7	3

Solution

Class boundaries	**Frequency (f)**	**Cumulative frequency (cf)**
10-15	4	4
15-20	12	16
20-25	17	33
25-30	23	56
30-40	11	67
40-50	9	76
50-60	7	83
60-70	3	86
	86	

P_{60} = The value of th item = The value of 51.6 th item which lies in the class 25-30

$$P_{60}=L+\frac{\frac{60N}{100}-cf}{f}\times I=25+\frac{51.6-33}{23}\times 5=25+4.04=29.04$$

P_{90} = The value of th item = The value of 77.4 th item which lies in the class 50-60

$$P_{90}=L+\frac{\frac{90N}{100}-cf}{f}\times I=50+\frac{77.4-76}{7}\times 10=50+2=52$$

From the decile and percentile results, we observed that $D_6 = P_{60}$ and $D_9 = P_{90}$.

4.5 Chapter Summary

This chapter has concisely introduced central tendencies (mean, median, mode) and partition values (quartiles, deciles, percentiles), emphasizing their practical applications in business data analysis. Through specific examples, it illustrates the optimal scenarios for employing each measure, enhancing the ability of business professionals to effectively interpret data and make informed decisions.

4.6 Self-Assessment

1. Define Arithmetic Mean, Geometric Mean, and Harmonic Mean and compare their advantages and disadvantages.
2. Proof any two positive numbers the relation will be: A.M. $\geq$ G.M. $\geq$ H.M.
3. Calculate the arithmetic mean and median of the following values:
   i) 12, 8, 9, 13, 15, 34, 45 ii) 16, 18, 22, 26, 29, 40
   iii) 67, 69, 86, 88, 94, 64, 68
4. Calculate the mode of the following values:
   i) 5, 8, 9, 8, 7, 10, 8, 15 ii) 15, 18, 19, 23, 24, 18,23, 22
   iii) 4, 5, 7, 8, 9, 12, 15, 19
5. Find the geometric mean and harmonic mean of the following numbers:
   i) 2, 4, 6, 8 ii) 3, 8, 10
   iii) 6, 12, 24
6. Determine the median from the following data:

Class	10-14	15-19	20-24	25-29	30-34	35-39	40-45
Frequency	23	45	66	72	50	32	12

7. Determine mean, median, and mode from the following data and explain their relations.

Wages	No of worker	Wages	No of worker
12.5-17.5	6	27.5-32.5	22
17.5-22.5	12	32.5-37.5	16
22.5-27.5	18	37.5-42.5	8

8. Calculate all the quartiles from the following data:

Class	20-29	30-39	40-49	50-59	60-69	70-79
Frequency	6	8	10	14	7	4

9. Find the median, 3rd quartile, 7th decile, and 60th percentile from the following data:

Age group	25-29	30-34	35-44	45-54	55-64	65-74
No of person (million)	3.1	4.2	2.8	6.9	4.9	1.8

Unit-V

Measures of Dispersion

5.1 Learning Objectives

By the end of this chapter, you will be able to:

i) Understand and apply Absolute Measures of Dispersion;

ii) Explore Relative Measures of Dispersion;

iii) Understand the importance of these measures in comparing the variability of data sets;

iv) Familiarize with the concept of moments in statistical distributions;

v) Learn about skewness and kurtosis as measures of distribution shape;

vi) Develop the ability to apply these concepts to real-world data.

5.2 Introduction

Nowadays, understanding the dispersion and shape of data distributions is as crucial as knowing their central tendency. This chapter, 'Measures of Dispersion and Shape of Distribution,' delves into the essential statistical tools that quantify the spread and shape of data sets. It covers key concepts such as range, variance, standard deviation, skewness, and kurtosis. These measures provide deeper insights into the consistency, reliability, and nature of business data, enabling professionals to make more nuanced and informed decisions. Through this chapter, readers will learn not just to describe data, but to understand the facts it tells about variability and distribution trends in a business context.

5.3 Measures of Dispersion

Measures of dispersion are statistical tools that quantify the spread or variability within a set of data. They include range, variance, standard deviation, and interquartile range. These measures are fundamental in statistics as they provide insight into the degree of variability or consistency in a dataset. Unlike measures of central tendency, which give a central value, dispersion measures the spread and extent of data around this central value.

Importance of Measures of Dispersion

i) **Understanding Data Variability:** Measures of dispersion are crucial for understanding how much individual data points differ from each other and from the average. This understanding is vital in fields like quality control, where consistency is key.

ii) **Enhancing Data Interpretation:** In combination with central tendency, dispersion measures give a complete picture of data. For instance, two data sets can have the same mean but vastly different ranges or variances, leading to different interpretations and decisions.

iii) **Risk Assessment:** In finance and investment, measures of dispersion like standard deviation are used to estimate the risk associated with an asset or portfolio. Higher variability often implies higher risk.

iv) **Comparing Data Sets:** Dispersion measures allow for the comparison of the spread of two or more different datasets, which is crucial in fields like market research and performance analysis.

v) **Identifying Outliers:** These measures help in identifying outliers which are data points significantly different from others. This is essential in data cleaning and ensuring the accuracy of statistical analysis.

Different Measures of Dispersion

Measures of dispersion are classified into two categories: absolute and relative. This classification is based on how the dispersion is expressed and the context in which it is used.

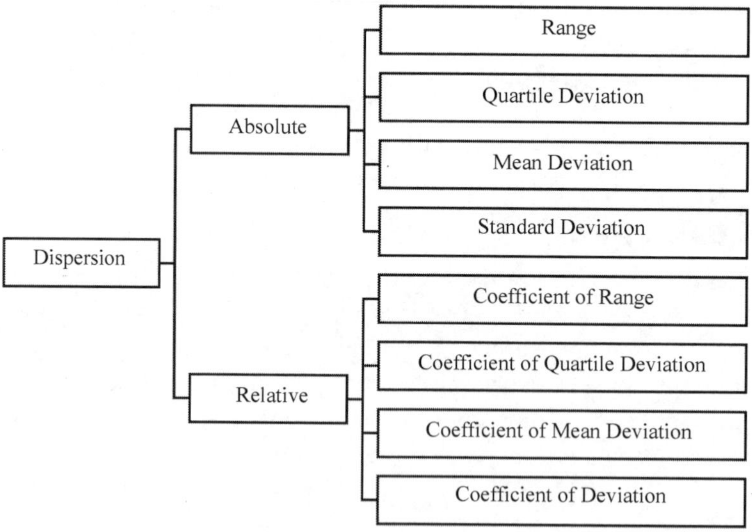

Absolute measures of dispersion provide a direct measure of the variability in a data set. They are expressed in the same units as the data itself. For instance, if you are measuring the variability in heights (in centimeters) of a group of individuals, an absolute measure like standard deviation would also be in centimeters. Absolute measures are particularly useful when comparing variability within the same dataset or datasets that have the same scale.

Relative measures of dispersion, on the other hand, are dimensionless. They provide a measure of variability relative to a central value (like the mean) and are expressed as a percentage, coefficient, or ratio. These measures are especially useful when comparing the variability of two or more datasets that are on different scales or have different units of measurement. For example, if comparing the variability of heights (in centimeters) and weights (in kilograms), a relative measure like the coefficient of variation allows for a meaningful comparison.

The choice between absolute and relative measures of dispersion often depends on the nature of the data and the specific requirements of the analysis. Absolute measures are more straightforward and intuitive but may not be suitable for comparisons across different units or scales. Relative measures, while providing a normalized form of variability, might require a deeper understanding of the data and its context. Both types play a crucial role in statistical analysis, offering different but complementary perspectives on the variability within data sets.

5.3.1 Range

The range is one of the simplest measures of dispersion in statistics. It is defined as the difference between the highest and lowest values in a data set. Mathematically, it is calculated as:

Range = Maximum Value - Minimum Value

When to Use Range

i) The range is most useful in data sets where the extreme values are of particular interest. It is often used in preliminary data analysis to get a quick sense of the spread in the data.

ii) It is also helpful in quality control and process monitoring contexts, where it is important to know the extent of variability.

Advantages of Range

i) **Simplicity and Ease of Understanding:** The range is straightforward to calculate and easy for most people to understand, making it a convenient tool for initial data exploration.

ii) **Quick to Compute:** Unlike more complex measures of dispersion, the range can be calculated rapidly, even for large data sets.

iii) **Useful for Identifying Outliers:** The range can help in quickly identifying outliers or extreme values in a data set, which can be important in certain analyses.

Disadvantages of Range

i) **Sensitivity to Extreme Values:** The range is highly sensitive to outliers or extreme values, which can give a misleading picture of the overall variability in the data.

ii) **Does Not Reflect Distribution of Data:** It does not provide any information about the distribution of values between the extremes, unlike measures like variance or standard deviation.

iii) **Not Useful for Open-End Classes:** It cannot be calculated from frequency distributions with open-end classes.

Calculation Procedure

Absolute Measure

Range = Maximum Value - Minimum Value

Example: Find range from the following data:

5, 25, 34, 49, 76, 24, 29, 80

Solution

$$\text{Relative range or coefficient of range} = \frac{\text{Range}}{\text{Sum of maximum and minimum values}}$$

Relative Measure

For a grouped frequency distribution, the range is calculated by the difference between the upper boundary of the highest class and the lower boundary of the lowest class. This is the ratio of the range to the median or mean of the dataset.

Example: Find the range and its coefficient from the following data:

Height (cm)	120-129	130-139	140-149	150-159	160-169
No of student	12	20	25	6	1

Range = 169.5 - 119.5 = 50 cm

$$\text{Ralative range} = \frac{50}{169.5+119.5} = \frac{50}{289}$$

Solution

In this case, we need to convert class limits to class boundaries to make it continuous. After that,

Range = 169.5 - 119.5 = 50 cm

$$\text{Relative range} = \frac{50}{169.5 + 119.5} = \frac{50}{289}$$

5.3.2 Quartile Deviation (Q.D.) or Semi-Interquartile Range

The quartile deviation is the measure of the spread of the middle 50% of a data set. It is defined as half the difference between the upper quartile (Q_3) and the lower quartile (Q_1).

Mathematically, it is expressed as:

$$\text{Q.D.} = \frac{Q_3 - Q_1}{2}$$

When to use Q.D.

i) The quartile deviation is particularly useful when you need a robust measure of dispersion that is not as sensitive to outliers as the range or standard deviation.

ii) It's often used in datasets with skewed distributions or when the data contains outliers.

Advantages of Q.D.

i) Robustness: Less sensitive to extreme values, making it more reliable for skewed distributions.

ii) Simplicity: Relatively easy to understand and compute.

iii) Focus on Middle Data: Emphasizes the spread of the middle half of the data, often where the most typical values lie.

iv) Appropriate for Open-End Classes: Q.D. is most favourable measure of dispersion for open-end grouped frequency distribution.

Disadvantages of Q.D.

i) Ignores Data Extremes: It does not take into account the variability present in the outer 50% of the data.

ii) Not as Commonly Used: Less familiar to some users compared to standard deviation or variance.

Calculation Procedure

Absolute Measure

$$Q.D. = \frac{Q_3 - Q_1}{2}$$

Relative Measure

$$\text{Coefficient of Q.D.} = \frac{\text{Quartile Deviation}}{\text{Median}} \times 100 = \frac{Q_3 - Q_1}{2Q_2} \times 100$$

Example: Find quartile and coefficient of quartile deviation from the following data:

Class	10-15	15-20	20-25	25-30	30-40	40-50	50-60	60-70
Frequency	4	12	17	23	11	9	7	3

Solution

Class boundaries	Frequency (f)	Cumulative frequency (cf)
10-15	4	4
15-20	12	16
20-25	17	33
25-30	23	56
30-40	11	67
40-50	9	76
50-60	7	83
60-70	3	86
	86	

Q_1 = The value of $\frac{iN}{4} = \frac{1}{4} \times 86$ th item = The value of 21.5 th item which lies in the class 20-25

$$Q_1 = L + \frac{\frac{N}{4} - of}{f} \times I = 20 + \frac{21.5 - 16}{17} \times 5 = 20 + 1.62 = 21.62$$

Median/Q_2= The value of th item = The value of 43 th item which lies in the class 25-30

$$Q_2 = L + \frac{\frac{2N}{4} - of}{f} \times I = 25 + \frac{43 - 33}{23} \times 5 = 25 + 2.17 = 27.17$$

Q_3= The value of $\frac{3N}{4} = \frac{3}{4} \times 86$ th item = The value of 64.5 th item which lies in the class 30-40

$$Q_3 = L + \frac{\frac{3N}{4} - \text{of}}{f} \times I = 30 + \frac{64.5-56}{11} \times 10 = 30 + 7.73 = 37.73$$

$$\text{Q.D.} = \frac{Q_3 - Q_1}{2} = \frac{37.73 - 21.62}{2} = 8.055$$

$$\text{Coel ficient of Q.D.} = \frac{\text{Q.D.}}{Q_2} \times 100 = \frac{8.055}{27.17} \times 100 = 29.65\%$$

5.3.3 Mean Deviation (M.D.)

The mean deviation, also known as the mean absolute deviation, is a measure of dispersion that indicates the average distance of all data points from a central point (typically the mean or median of the dataset).

Mathematically, it can be expressed as:

$$\text{M.D.} = \frac{1}{N} \sum |x_i - \bar{x}|$$

When to Use M.D.

i) Mean deviation is particularly useful for understanding the average variability in a data set. It is most effective when the data set does not contain extreme outliers.

ii) It is often used in fields like quality control, economics, and social science research to gauge the consistency and reliability of data.

Advantages of M.D.

i) Intuitive Understanding: Provides a straightforward and easy-to-understand indication of dispersion around the mean or median.

ii) Balanced View of Spread: Offers a balanced view by considering the deviation of all points from the central value, rather than focusing only on the extremes.

iii) Less Sensitive to Extreme Values: Compared to the standard deviation, mean deviation is not squared, making it less sensitive to extreme values.

Disadvantages of M.D.

i) Not as Commonly Used: Mean deviation is not as widely used or recognized as other measures like standard deviation, which can make its interpretation less intuitive in some contexts.

ii) Less Mathematical Convenience: In statistical analysis, the mean deviation is often overshadowed by the standard deviation because it lacks certain mathematical properties that make the standard deviation more convenient for complex analyses.

iii) **Potential for Misinterpretation:** Without a clear understanding, the mean deviation can sometimes lead to misinterpretation of data variability, especially in highly skewed distributions.

Calculation Procedure

Absolute Measure

For unweighted data, M.D. is:

$$\text{M.D.} = \frac{1}{n}\sum_{i=1}^{n}|x_i - A|$$

Where, n = number of observation and A is the mean or median or any specified value.

For weighted data,

$$\text{M.D.} = \frac{1}{N}\sum_{i=1}^{n}f_i|x_i - A|$$

Where, $N = \Sigma f_i$ and A is the mean or median or any specified value.

Mean deviation is mainly two types based upon the uses of mean or median function. Two types are: **(i)** Mean deviation about mean, and **(ii)** Mena deviation about median.

Relative Measure

$$\text{Coef ficient of M.D.} = \frac{\text{M.D. about Mean}}{\text{Mean}} \times 100$$

Or,

$$\text{Coef ficient of M.D.} = \frac{\text{M.D. about Median}}{\text{Median}} \times 100$$

Example 1: Find the mean deviation about median and coefficient from the following data:

22, 15, 18, 30, 35, 65, 47, 78, 67

Solution

We first arrange the numbers in ascending order 15, 18, 22, 30, 35, 47, 65, 67, 78

The number of observations (N) = 9

$$\text{Median} = \frac{N+1}{2}\text{th item} = \frac{9+1}{2} = 5\text{th item} = 35$$

x	$\lvert x_i - 35\rvert$; A=35
15	20
18	17
22	13
30	5
35	0
47	12
65	30
67	32
78	43
	172

$$\text{M.D.} = \frac{1}{n}\sum_{i=1}^{n}\lvert x_i - A\rvert = \frac{1}{9}\times 172 = 19.11$$

$$\text{Coel ficient of M.D.} = \frac{19.11}{35}\times 100 = 54.6\%$$

Example 2: Calculate mean deviation about mean and its coefficients from the following data:

Value	1	2	3	4	5	6	7	8
Frequency	5	8	7	11	17	20	14	2

Solution

Value (x)	Frequency (f)	*fx*	$\lvert x_i - \overline{x}\rvert$	$f\lvert x_i - \overline{x}\rvert$
1	5	5	3.82	19.1
2	8	16	2.82	22.56
3	7	21	1.82	12.74
4	11	44	0.82	9.02
5	17	85	0.18	3.06
6	20	120	1.18	23.6
7	14	98	2.18	30.52
8	2	16	3.18	6.36
Total	84	405		126.96

$$\overline{x} = \frac{\sum fx}{N} = \frac{405}{84} = 4.82$$

$$\text{M.D. about mean} = \frac{1}{N}\sum_{i=1}^{n} f_i |x_i - \bar{x}| = \frac{1}{84} \times 126.96 = 1.51$$

$$\text{Coel ficient of M.D.} = \frac{\text{M.D. about Mean}}{\text{Mean}} \times 100 = \frac{1.51}{4.82} \times 100 = 31.33\%$$

Example 3: Compute mean deviation about median and its coefficients

Weight (kg)	31-35	36-40	41-45	46-50	51-55	56-60	61-65
No of Person	10	22	32	45	50	43	26

Solution

Class limit	Class boundaries	Frequency (f)	*cf*	Mid value (x)	$\|x_i - Me\|$	f $\|x_i - Me\|$
31-35	30.5-35.5	10	10	33	18.8	188
36-40	35.5-40.5	22	32	38	13.8	303.6
41-45	40.5-45.5	32	64	43	8.8	281.6
46-50	45.5-50.5	45	109	48	3.8	171
51-55	50.5-55.5	50	159	53	1.2	60
56-60	55.5-60.5	43	202	58	6.2	266.6
61-65	60.5-65.5	26	228	63	11.2	291.2
		228 (N)				1562

Median class = The value of N/2th item

= The value of 114 (228/2) th item which lies in the class (50.5-55.5)

$$\text{Median} = L + \frac{\frac{N}{2} - cf}{f} \times I = 50.5 + \frac{114 - 109}{50} \times 5$$

$$= 50.5 + 1.3 = 51.8\,\text{kg}$$

$$\text{M.D. about median} = \frac{1}{N}\sum_{i=1}^{n} f_i |x_i - A| = \frac{1}{228} \times 1562 = 6.85$$

$$\text{Coef ficient of M.D.} = \frac{6.85}{51.8} \times 100 = 13.22\%$$

The Coefficient of Mean Deviation of 13.22% indicates that the average deviation of the weights from the median (51.8) is 13.22% of the median.

5.3.4 Standard Deviation (S.D.)

Standard deviation is a widely used measure of variability or dispersion in statistics. It quantifies the amount of variation or dispersion of a set of data values. A low standard deviation indicates that the data points tend to be close to the mean (or expected value) of the set, while a high standard deviation indicates that the data points are spread out over a wider range. SD may be defined as "Root-Mean-Square-Deviation from mean." Mathematically, it can be expressed as:

$$SD(\sigma) = \sqrt{\frac{1}{n}\sum_{i=1}^{n}(x_i - \bar{x})2}$$

When to Use SD

Standard deviation is particularly useful in datasets where you need to understand the extent of variation from the average. It is extensively used in fields like finance for risk assessment, in quality control processes, and in scientific research to quantify the degree of uncertainty in experimental results.

Advantages of SD

i) **Clarity and Precision:** Provides a clear and precise measure of how spread out numbers in a dataset are.

ii) **Consistency with the Mean:** It is based on the mean and gives a measure of variability in relation to that mean.

iii) **Foundation for Further Statistical Analysis:** It is used in various statistical methods, including confidence intervals and hypothesis testing.

Disadvantages of SD

i) **Not Useful for Comparing Dispersions in Different Units:** The dispersions of two or more series of observations provided in different units cannot be compared using this method.

ii) **Squared Units:** The standard deviation squares the units of the original data, which can sometimes make interpretation less intuitive.

iii) **Assumption of Normality:** The standard deviation is most effective for data that is normally distributed; its interpretation can be more complex for skewed distributions.

Calculation Procedure

Absolute Measure

For unweighted data, SD formula is:

$$SD(\sigma) = \sqrt{\frac{1}{n}\sum_{i=1}^{n}(x_i - \bar{x})^2}$$

Or, the working formula is as follows:

$$\sigma = \sqrt{\frac{\sum x^2}{n} - \left(\frac{\sum x}{n}\right)^2}$$

For weighted data, SD formula is:

$$SD(\sigma) = \sqrt{\frac{1}{N}\sum_{i=1}^{n} f_i (x_i - \bar{x})^2}$$

Where, $\Sigma^{i}_{n=1} f_i$

Working formula is as follows:

$$\sigma = \sqrt{\frac{\sum fx^2}{N} - \left(\frac{\sum fx}{N}\right)^2}$$

SD formula based on step division method is as follows:

$$\sigma = \sqrt{\frac{\sum fd^2}{N} - \left(\frac{\sum fd}{N}\right)^2} \times I$$

Where, $d_i \frac{x_i - A}{I}$;i= 1, 2, 3, ... n; A= change in origin and I= change in scale (class interval).

Relative Measure

$$\text{Coef ficient of var iation (C.V.)} = \frac{SD}{Mean} \times 100$$

Example 1: Compute standard deviation and CV from the following data:

x	20	30	40	50	60	70	Total
f	10	17	25	27	15	6	100

Solution using actual mean method

x	*f*	*fx*	*fx*²
20	10	200	4000
30	17	510	15300
40	25	1000	40000
50	27	1350	67500
60	15	900	54000
70	6	420	29400
Total	100	4380	210200

$$\sigma=\sqrt{\frac{\sum fx^2}{N}-\left(\frac{\sum fx}{N}\right)^2}$$

$$\sigma=\sqrt{\frac{210200}{100}-\left(\frac{4380}{100}\right)^2}=\sqrt{2102-1918.44}=13.55$$

Solution using step deviation method

x	*f*	$d=\frac{x-1}{I}$ A=50; I=10	*fd*	*fd*²
20	10	-3	-30	90
30	17	-2	-34	68
40	25	-1	-25	25
50	27	0	0	0
60	15	1	15	15
70	6	2	12	24
Total	100		-62	222

$$\sigma=\sqrt{\frac{\sum fd2}{N}-\left(\frac{\sum fd}{N}\right)^2}\times I$$

$$\sigma=\sqrt{\frac{222}{100}-\left(\frac{-62}{100}\right)^2}\times 100$$

$$\sigma=\sqrt{2.22-0.3844}\times 10=13.55$$

$$C.V.=\frac{13.55}{43.80}\times 100=30.94\%$$

[Where, $=\frac{\sum fx}{N}=\frac{4380}{100}=43.80$]

Example 2: Compute standard deviation and coefficient of variation from the following data:

Marks	20-29	30-39	40-49	50-59	60-69	70-79	80-89
No of student	6	16	17	25	15	12	9

Solution based on actual mean method

Class limit	Class boundaries	f	Mid value (x)	fx	fx^2
20-29	19.5-29.5	6	24.5	147	3601.5
30-39	29.5-39.5	16	34.5	552	19044
40-49	39.5-49.5	17	44.5	756.5	33664.25
50-59	49.5-59.5	25	54.5	1362.5	74256.25
60-69	59.5-69.5	15	64.5	967.5	62403.75
70-79	69.5-79.5	12	74.5	894	66603
80-89	79.5-89.5	9	84.5	760.5	64262.25
Total		100		5440	323835

$$\sigma=\sqrt{\frac{\sum fx2}{N}-\left(\frac{\sum fx}{N}\right)^2}$$

$$\sigma=\sqrt{\frac{323835}{100}-\left(\frac{5440}{100}\right)^2}=\sqrt{3238.35-2959.36}=16.70$$

$$C.V.=\frac{16.70}{54.40}\times 100=30.70\%$$

$$[\text{Where Mean}=\frac{\sum fx}{N}=\frac{5440}{100}=54.40]$$

Solution based on Step Deviation Method

Class limit	Class boundaries	f	Mid value (x)	$d=\frac{x-A}{I}$ **A=54.5; I=10**	fd	fd^2
20-29	19.5-29.5	6	24.5	-3	-18	54
30-39	29.5-39.5	16	34.5	-2	-32	64
40-49	39.5-49.5	17	44.5	-1	-17	17
50-59	49.5-59.5	25	54.5	0	0	0
60-69	59.5-69.5	15	64.5	1	15	15
70-79	69.5-79.5	12	74.5	2	24	48
80-89	79.5-89.5	9	84.5	3	27	81
Total		100			-1	279

$$\sigma=\sqrt{\frac{279}{100}-\left(\frac{-1}{100}\right)^2}\times 10$$

$$\sigma=\sqrt{2.79-0.0001}\times 10=16.70$$

$$\text{A.M.}(\bar{x})=A+\frac{\sum fd}{N}\times I$$

$$x=54.5+\frac{-1}{100}\times 10$$

$$=54.5-0.1=54.4$$

$$\text{C.V.}=\frac{16.70}{54.4}\times 100=30.70\%$$

Combined Standard Deviation

Let there be two groups having respectively the number of observations n_1 and n_2; the arithmetic means and , the standard deviations and. These two groups are combined together to form a new group. The standard deviation of this combined group is given by

$$\sigma=\sqrt{\frac{n_1(\sigma_1^2+d_1^2)+n_2(\sigma_2^2+d_2^2)}{n_1+n_2}}$$

$$\text{Where } d_1=\bar{x}_1-\bar{x}, d_2=\bar{x}_2-\bar{x} \text{ and } \bar{x}=\frac{n_1 x_1+n_2 x_2}{n_1+n_2}$$

For three groups the corresponding formula for the combined SD is given by

$$\sigma=\sqrt{\frac{n_1(\sigma_1^2+d_1^2)+n_2(\sigma_2^2+d_2^2)+n_3(\sigma_3^2+d_3^2)}{n_1+n_2+n_3}}$$

Example: Compute combined mean and SD from the following data:

Category	Sample size	Mean	SD
A	50	54.4	8
B	100	50.3	7

Solution

$$\bar{x}=\frac{n_1\bar{x}_1+n_2\bar{x}_2}{n_1+n_2}=\frac{(50\times 54.4)+(100\times 50.3)}{50+100}=51.67$$

$$\text{Combined SD}(\sigma)=\sqrt{\frac{n_1(\sigma_1^2+d_1^2)+n_2(\sigma_2^2+d_2^2)}{n_1+n_2}}$$

Here, $d_1 = \bar{x}_1 - \bar{x} = 54.4 - 51.67 = 2.73, d_2 = x_2 - \bar{x} = 50.3 - 51.67 = -1.37$,

$$\sigma = \sqrt{\frac{50(8^2 + (2.73)^2) + 100(7^2 + (-1.37)^2)}{50 + 100}}$$

$$\sigma = \sqrt{\frac{50(64 + 7.453) + 100(49 + 1.877)}{150}}$$

$$\sigma = \sqrt{\frac{3572.65 + 5087.7}{150}}$$

Some Important Properties of SD

i) SD is independent of the change of origin but dependent on change in scale.

Let a variable x assumes n different values with respective frequencies. Then, from the definition of standard deviation we have

$$\sigma x^2 = \frac{1}{N}\sum f(x - \bar{x})^2 \qquad \ldots (1)$$

Let us now change the origin and scale of x respectively by A and I, so that the new variable becomes d.

Thus, $d = \frac{x - A}{I}$ or, $x = A + dl$

$\therefore x = A + \bar{d}l$

Now, $x - \bar{x} = (A + dl) - (A + \bar{d}l) = (d - \bar{d})I$

Therefore, from (1), we get

$$\sigma x^2 = \frac{1}{N}\sum f(d - \bar{d})^2 . I^2 = I^2 . \frac{1}{N}\sum f(d - \bar{d})^2 = I^2 \sigma_d^2$$

$\therefore \sigma_x = |I| \sigma d$

Thus, we find that the final result does not contain the term A, but it contains the term I. Hence, standard deviation is independent on origin but depends on scale.

ii) SD is the minimum root-mean-square-deviation.

Let a variable x assumes n different values with respective frequencies. Then, from the definition of standard deviation we have

$$SD(\sigma) = \sqrt{\frac{1}{n}\sum_{i=1}^{n} (xi - x)^2} \qquad \ldots (1)$$

If instead of taking deviations from mean, we consider deviations from an arbitrary constant A, then

$$\text{Root}-\text{Mean}-\text{Square deviation from A} = \sqrt{\sum (x_i - A)^2 / n} \qquad \ldots (2)$$

We have to show that (2) cannot be smaller than (1) whatever be the value of A.

Now, it was already proved that for a given set of observations the sum of the squares of deviation is the minimum, when deviations are taken from the arithmetic mean [see note]. So that, we get

$$\sum (x_i - A)^2 = \sum (x_i - x)^2 + n(\overline{x} - A)^2$$

Since the second term on the right can never be negative

$$\sum (x_i - A)^2 = \sum (x_i - \overline{x})^2$$

Dividing both sides by n and taking the square-root, we have

$$\sqrt{\sum (x_i - A)^2 / n} \geq \sqrt{\sum (x_i - \overline{x})^2 / n}$$

Note: Prove that for a given set of observations the sum of the squares of deviation is the minimum, when deviations are taken from the arithmetic mean.

From the problem statement, we need to prove if $\Sigma(x_i - A)^2$, then $\Sigma(x_i - A)^2/n$ will be the least.

Now, we can write:

$$x_i - A = (x_i - \overline{x}) + (\overline{x} - A)$$

Therefore,

$$\sum (x_i - A)^2 = \sum \{(x_i - \overline{x}) + (\overline{x} - A)\}^2$$

$$= \sum \{(x_i - \overline{x})^2 + 2(x_i - \overline{x})(\overline{x} - A) + (\overline{x} - A)\}^2$$

$$= \sum (x_i - \overline{x})^2 + \sum 2(x_i - \overline{x})(\overline{x} - A) + \sum (\overline{x} - A)^2$$

$$= \sum (x_i - \overline{x})^2 \, 2(\overline{x} - A) \sum (x_i - \overline{x}) + n(\overline{x} - A)^2$$

$$= \sum (x_i - \overline{x})^2 + 2(\overline{x} - A).0 + n(\overline{x} - A)^2$$

i.e.

$$\sum (x_i - A)^2 = \sum (x_i - \overline{x})^2 + n(\overline{x} - A)^2$$

Both expressions on the right side are inherently positive. This is because the first expression represents the total of n squared differences $(x_i - \overline{x})^2$, and the second is the multiplication of n (a positive integer) by a squared term. Our goal, however, is to find a specific value for A that minimizes the squared differences $(x_i - A)^2$. To achieve the lowest possible value for these differences, we aim for the second expression on the right side to be as close to zero as possible, that is, when A is chosen such that the squared differences are minimized.

$$n(\overline{x} - A)^2 = 0$$

Or, $(x - A)^2 = 0$

Or, $(\overline{x} - A) = 0$

$\therefore A = \overline{x}$

Relation between SD and Other Absolute Measures

For a normal to moderately skewed distribution, following empirical relations hold:

A. Range= 6 SD

B. Mean Deviation= $\frac{4}{5}$ SD

C. Quartile Deviation= $\frac{2}{3}$ SD

5.3.5: Lorenz Curve

The Lorenz curve is a graphical representation used to demonstrate the distribution within a group. Essentially, it's a graph that accumulates percentages, linking the proportion of the population being analyzed to the proportion of a specific characteristic within that group. Taking wealth distribution as an example, if this wealth was evenly spread across individuals, the curve would form a straight line from one end of the scale to the other (ACB). In reality, however, the curve usually takes a shape more like ADB. The smaller the gap between the Lorenz curve (ADB) and the straight diagonal line (ACB), the more evenly distributed the wealth is, indicating less variability. Conversely, a larger gap suggests a higher concentration of wealth among fewer individuals, signifying more disparity. Unlike other statistical measures such as range or standard deviation, the Lorenz curve doesn't provide a precise numerical value of dispersion. Its shortcoming is that it's less precise than these conventional measures. However, its visual nature allows for an immediate understanding of the distribution. The Lorenz curve is particularly valuable for examining the distribution of resources like land, income, or profits across various segments of a nation's population or among different business groups.

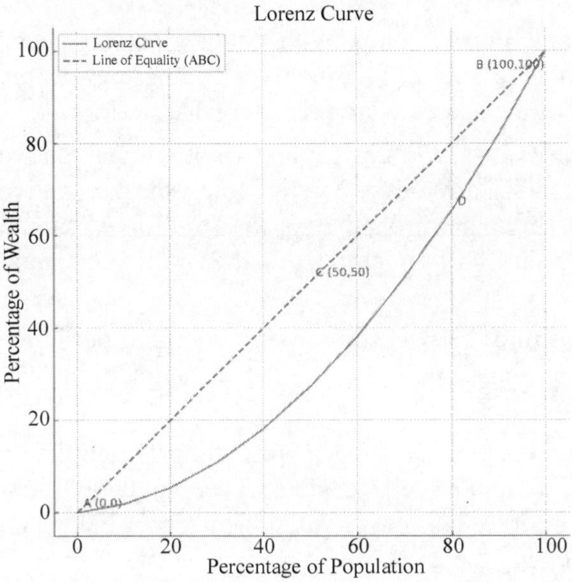

Lorenz Curve Construction Steps

To construct a Lorenz curve, which illustrates the distribution of a variable (such as income) across a population, follow these steps:

Step 1: Calculate Percentage Frequency: For each income class, determine the percentage frequency. This is done by dividing the frequency of each income class (from your data) by the total frequency of all classes and then multiplying by 100. This calculation gives you the proportion of the population in each income class.

Step 2: Cumulative Percentage Frequency: Calculate the cumulative percentage for these frequencies. This means adding the percentage frequency of each class to the sum of the frequencies of all preceding classes.

Step 3: Determine Class Mid-Points: Compute the mid-point for each income class. This is often the average income for each class and can be seen as a representative income level for the class.

Step 4: Total Income of Each Class: To understand the share of each income class, you need to know its total income. This is found by multiplying the frequency of each class (number of people) by the class mid-point (average income).

Step 5: Calculate Percentage Share of Income: For each class, calculate the percentage share of the total income. This is done by dividing the total income of each class (from the previous step) by the total income of all classes, then multiplying by 100.

Step 6: Cumulative Percentage Share: Compute the cumulative percentage for these income shares. Similar to step 2, this involves adding the income share of each class to the sum of the shares of all preceding classes.

Step 7: Graphing the Lorenz Curve: On graph paper, plot the cumulative percentage of the variable (income share) on the Y-axis against the cumulative percentage of the population (frequency) on the X-axis. Both axes should range from 0% to 100%. The point (0,0) represents that 0% of the population holds 0% of the income.

Step 8: Line of Equal Distribution: Draw a straight line from the coordinate (0,0) to (100,100). This line represents a perfectly equal distribution of income across the population.

Step 9: Plotting the Lorenz Curve: Plot the points representing the cumulative percentage of income against the cumulative percentage of the population. Join these points to form the Lorenz curve. This curve will typically bow below the line of equal distribution, indicating the level of income inequality.

Example: Create a Lorenz Curve based on the following data of the monthly income of employees of a company.

Income	0-5000	5000-10000	10000-20000	20000-40000	40000-50000
No of employee	5	10	18	10	7

Solution

Class	**Frequency (f)**	**Percentage frequency**	**Cumulative percentage frequency**	**Mid value**	**Total income**	**Percentage share of income**	**Cumulative percentage share of income**
0-5000	5	10	10	2500	12500	1.29	1.29
5000-10000	10	20	30	7500	75000	7.71	9
10000-20000	18	36	66	15000	270000	27.76	36.76
20000-40000	10	20	86	30000	300000	30.85	67.61
40000-50000	7	14	100	45000	315000	32.39	100
	50	100			972500	100	

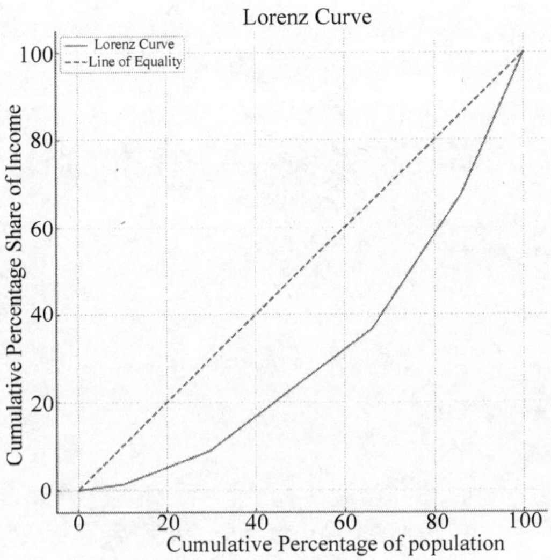

5.4 Shape of Distribution

In statistics, understanding the shape of a distribution is as crucial as understanding its central tendency and dispersion. The shape, characterized by aspects like moments, skewness, and kurtosis, offers profound insights into the nature of the data. It helps in identifying the symmetry or asymmetry of the distribution, thus revealing any potential biases in the data. Skewness measures the degree of asymmetry, while kurtosis indicates the 'tailedness' of the distribution, providing clues about the likelihood of outliers. These measures are invaluable in various fields, from finance to social sciences, where they assist in making informed decisions and predictions based on the underlying data patterns. Understanding the shape of distribution is, therefore, a key step in the comprehensive analysis of data.

5.4.1 Moments

The rth moment of a variable x about a given number A, usually denoted is defined as

$$m'_r = \frac{1}{N}\sum_{i=1}^{n} f_i (x_i - A)^r \qquad \dots (1)$$

For instance, the 3rd moment (r=3) about 8 (A=8) will be given by $m'_3 = \frac{1}{N}\sum_{i=1}^{n} f_i (x_i - 8)^3$.

For unweighted data, the corresponding formula for the moment is obtained by putting in the previous formula. Thus, we get

$$m_r' = \frac{1}{n}\sum(x_i - A)^r$$

For a frequency distribution, two particular cases are

Case I: When A=0, the moment is known as the raw moment. Thus, from (1), the rth raw moment is:

$$m_r' = \frac{1}{N}\sum_{i=1}^{n} f_i (x_i - 0)^r = \frac{1}{N}\sum_{i=1}^{n} f_i x_i^r$$

So, 1st raw moment will be: $m_1' = \frac{1}{N}\sum_{i=1}^{n} f_i x_i = \text{Mean}$

2nd raw moment will be: $m_2' = \frac{1}{N}\sum_{i=1}^{n} f_i x_i^2$

Case II: When $A = \bar{x}$ (i.e., arithmetic mean), the moment is known as the central moment and is denoted by .

[*Note that moments about mean are written without dashes ('), but moments about any other origin, i.e. non central moments, with dashes*.]

$$m_r = \frac{1}{N}\sum f_i (x_i - \bar{x})^r$$

This is obtained from (1) by putting.

Thus, $m_1 = \frac{1}{N}\sum f_i (x_i - \bar{x}) = \frac{1}{N}\sum f_i x_i - \frac{1}{N}\sum f_i \bar{x} = \bar{x} - \bar{x}.\frac{1}{N}\sum f_i = \bar{x} - \bar{x} = 0 [\text{since } N = \sum f_i]$

$$m_2 = \frac{1}{N}\sum f_i (x_i - x)2 = (S.D)^2 = \text{Variance}$$

$$m_3 = \frac{1}{N}\sum f_i (x_i - \bar{x})^3$$

Similarly, the other central moments can be obtained.

Example: calculate first three central moments from the following distribution.

x	5	7	10	12	17
f	1	2	4	2	1

Solution

x	f	fx	$x - \bar{x}$	$f(x - \bar{x})$	$f(x - \bar{x})^2$	$f(x - \bar{x})^3$
5	1	5	-5	-5	25	-125
7	2	14	-3	-6	18	-54

10	4	40	0	0	0	0
12	2	24	2	4	8	16
17	1	17	7	7	49	343
Total		100		0	100	180

$$x = \frac{1}{N}\sum fx = \frac{1}{10}\times 100 = 10$$

$$m1 = \frac{1}{N}\sum f(x - x) = \frac{1}{10}\times 0 = 0$$

$$m_2 = \frac{1}{N}\sum f_i(x_i - \overline{x})^2 = \frac{1}{10}\times 100 = 10$$

$$m_2 = \frac{1}{N}\sum f_i(x_i - \overline{x})^3 = \frac{1}{10}\times 180 = 18$$

Beta-coefficients and Gamma-coefficients

Beta and Gamma are two Greek small letters. There are two beta coefficients

$$\beta_1 = \frac{{m_3}^2}{{m_3}^2}; \beta_2 = \frac{m_4}{{m_2}^2}$$

The beta coefficients are never negative and are pure numbers, independent of origin and scale of observations.

There are two Gamma-coefficients, namely,

$$\gamma_1 = \sqrt{\beta_1} = \sqrt{\frac{m_4}{m_2}} = \frac{m^3}{\sigma^2} \text{ and } \gamma_2 = \beta_2 - 3 = \frac{m_4}{m_2} - 3$$

Gamma-coefficients may be positive, negative, or zero. They are also pure numbers like Beta-coefficients.

5.4.2 Skewness

A distribution of frequency is considered symmetrical when values at equal distances on either side of the mean exhibit the same frequency. 'Skewness' refers to the degree of asymmetry present in the data. If a frequency distribution deviates from being symmetrical, it is described as 'skewed'. Essentially, 'skewness' implies a departure from symmetry, with 'skew' indicating a distribution that is not symmetrical. Consequently, a distribution that is perfectly symmetrical would have a skewness of zero. Depending on the direction of the asymmetry, skewness can be either positive or negative.

Skewness is calculated using the formula:

1. Pearson's First Measure

$$\text{Skewness} = \frac{\text{Mean - Mode}}{\text{Standard Deviation}}$$

2. Pearson's Second Measure

$$\text{Skewness} = \frac{3(\text{Mean} - \text{Median})}{\text{Stadard Deviation}}$$

3. Bowley's Measure

$$\text{Skewness} = \frac{(Q_3 - Q_2) - (Q_2 - Q_1)}{(Q_3 - Q_2) + (Q_2 - Q_1)}$$

$$= \frac{Q_3 - 2Q_2 + Q_1}{Q_3 \, Q_1}$$

Where Q_1, Q_2, Q_3 represent the first, second, and third quartiles of the distribution.

Moment Measure

$$\text{Skewness}(\gamma_1) = \frac{m_3}{\sigma_3} = \frac{m_3}{(\sqrt{m_2})}$$

In this context, and represent the second and third central moments, respectively, while is the standard deviation (S.D.).

Based on the skewness value, distributions can be classified into three distinct types:

a) **Positively Skewed Distribution (Skewness >0):** In such distributions, the Mean exceeds the Median, which in turn is greater than the Mode (Mean > Median > Mode). These distributions feature a frequency curve that extends with a longer tail towards the right side.

b) **Negatively Skewed Distribution (Skewness <0):** These distributions are characterized by a Mean that is less than the Median, which is further less than the Mode (Mean < Median < Mode). The frequency curve in this case exhibits a longer tail extending towards the left.

c) **Symmetrical Distribution (Skewness=0):** For distributions that are symmetrical, the Mean, Median, and Mode are all equal (Mean = Median= Mode). Additionally, the second quartile (Q_2) is the midpoint of the first (Q_1) and the third quartiles (Q_3) or $Q_2 = \frac{1}{2}(Q_1 + Q_3)$.

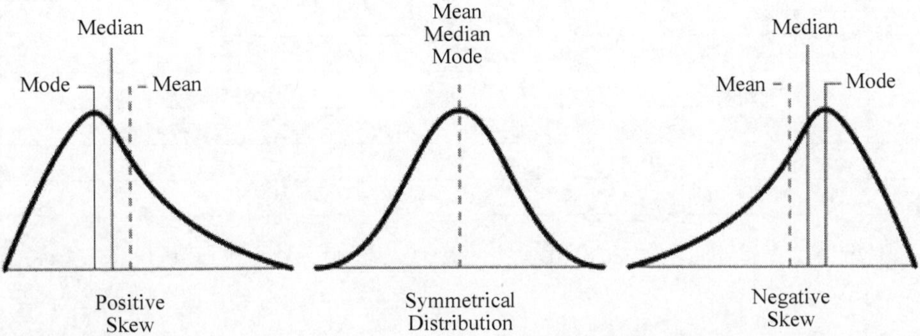

Example 1: Compute skewness value when $Q_1 = 30$, $Q_2 = 40$, and $Q_3 = 50$.

Solution

Using Bowley's measure

$$Sk = \frac{Q_3 - 2Q_2 + Q_1}{Q_3 - Q_1} = \frac{50-(2\times40)+30}{50-30} = \frac{80-80}{20} = 0$$

Example 2: Compute skewness value while the second and third central moments of a distribution is 4 and 12 respectively.

By using moment measure technique, we get

$$\text{Skewness}(\gamma_1) = \sqrt{\frac{m_3^2}{m_2^3}} = \sqrt{\frac{12^2}{4^3}} = \sqrt{\frac{12\times12}{4\times4\times4}} = \sqrt{\frac{9}{4}} = \frac{3}{2} = 1.5$$

Example 3: For a moderately skewed distribution, Mean = 175, Median = 165, SD = 40. Find the skewness of the distribution.

By using Pearson second measure, we get

$$\text{Skewness} = \frac{3(\text{Mean - Median})}{\text{Standard Deviation}} = \frac{3(175\text{-}165)}{40} = \frac{30}{40} = 0.75$$

Example 4: Compute skewness based on the Bowley's method.

Class	10-15	15-20	20-25	25-30	30-40	40-50	50-60	60-70
Frequency	4	12	17	23	11	9	7	3

Solution

Class boundaries	Frequency (f)	Cumulative frequency (cf)
10-15	4	4
15-20	12	16
20-25	17	33

25-30	23	56
30-40	11	67
40-50	9	76
50-60	7	83
60-70	3	86
	86	

Q_1 = The value of $\frac{iN}{4}=\frac{1}{4}\times 86$ th item = The value of 21.5 th item which lies in the class 20-25

$$Q_1 = L+\frac{\frac{N}{4}-cf}{f}\times I = 20+\frac{21.5-16}{17}\times 5 = 20+1.62 = 21.62$$

Q_2= The value of $\frac{2N}{4}=\frac{2}{4}\times 86$ th item = The value of 43 th item which lies in the class 25-30

$$Q_2 = L+\frac{\frac{2N}{4}cf}{f}\times I = 25+\frac{43-33}{23}\times 5 = 25+2.17$$

Q_3= The value of $\frac{3N}{4}=\frac{3}{4}\times 86$ th item = The value of 64.5 th item which lies in the class 30-40

$$Q_3 = L+\frac{\frac{3N}{4}-cf}{f}\times I = 30+\frac{64.5-56}{11}\times 10 = 30+7.73 = 37.73$$

$$Sk = \frac{Q_3-2Q_2+Q_1}{Q_3-Q_1} = \frac{37.73-(2\times 27.17)+21.62}{37.73-21.62} = \frac{5.01}{16.11} = 0.31$$

Example 5: Compute skewness value based on the Pearson's first and second measures from the following data.

Marks	20-29	30-39	40-49	50-59	60-69	70-79	80-89
No of student	6	16	17	25	15	12	8

Solution

Class limit	Class boundaries	f	Mid value (x)	fx	fx²	Cum. f
20-29	19.5-29.5	6	24.5	147	3601.5	6
30-39	29.5-39.5	16	34.5	552	19044	22
40-49	39.5-49.5	17	44.5	756.5	33664.25	39
50-59	49.5-59.5	25	54.5	1362.5	74256.25	64

60-69	59.5-69.5	15	64.5	967.5	62403.75	79
70-79	69.5-79.5	12	74.5	894	66603	91
80-89	79.5-89.5	8	84.5	676	57122	99
Total		99		5355.5	316694.75	

$$\text{Mean} = \frac{\sum fx}{N} = \frac{5355.5}{99} = 54.10$$

Median = The value of (99/2)th i.e. 49.5[th] item which lies in the class 49.5-59.5.

$$\text{Median} = L + \frac{\frac{N}{2} - cf}{f} \times I = 49.5 + \frac{49.5 - 39}{25} \times 10 = 53.7$$

$$\text{Mode} = l_1 + \frac{d_1}{d_1 + d_2} \times c = 49.5 + \frac{(25-17}{(25-17)+(25-15)} \times 10 = 49.5 + \frac{8}{18} \times 10 = 53.94$$

$$SD(\sigma) = \sqrt{\frac{\sum fx^2}{N} - \left(\frac{\sum fx}{N}\right)^2} = \sqrt{\frac{316694.75}{99} - \left(\frac{5355.5}{99}\right)^2} = \sqrt{3198.94 - 2926.81} = 16.50$$

Pearson's First Measure

$$\text{Skewness} = \frac{\text{Mean - Mode}}{\text{Standard Deviation}} = \frac{54.10 - 53.94}{16.50} = 0.01$$

Pearson's Second Measure

$$\text{Skewness} = \frac{3(\text{Mean - Median})}{\text{Standard Deviation}} = \frac{3(54.10 - 53.7)}{16.50} = 0.07$$

5.4.3 Kurtosis

Kurtosis, derived from a Greek term, signifies the sharpness or flatness of a frequency distribution's peak. It's possible for two distributions to be similar in terms of central tendency, dispersion, and skewness, yet differ in how closely their values concentrate around the mode. This means one distribution's frequency curve might be more sharply peaked compared to another. This attribute of a frequency distribution is referred to as kurtosis.

Kurtosis is calculated using the formula:

$$\text{Kurtosis} = \gamma_2 = \beta_2 = \frac{m_4}{(m_2)^2} - 3$$

This value is a dimensionless quantity, unaffected by changes in origin or scale. Based on the kurtosis value, distributions are categorized into three types:

i) Platy-kurtic distribution ($\gamma_2 <$ i.e., $\beta_2 < 3$)

ii) Meso-kurtic distribution (γ_2 i.e., $\beta_2 = 3$)

iii) Lepto-kurtic distribution (γ_2 i.e., $\beta_2 > 3$)

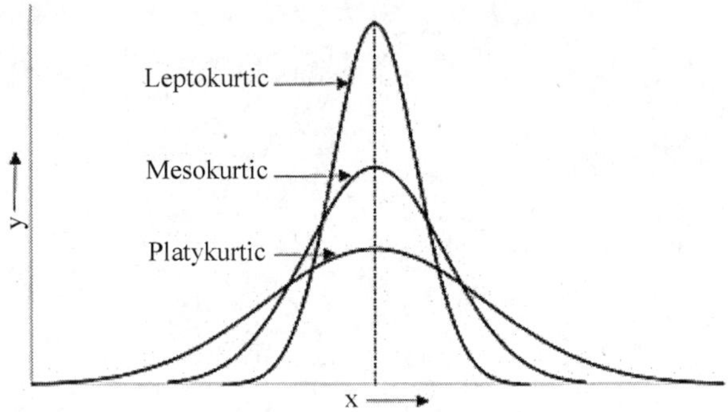

Example: Compute skewness and kurtosis from the following distribution using moment method.

Class	0-10	10-20	20-30	30-40
Frequency	1	3	5	2

Solution

Class	f	x	d=(x-25)/10	*fd*	*fd²*	*fd²*	*fd⁴*
0-10	1	5	-2	-2	4	-8	16
10-20	3	15	-1	-3	3	-3	3
20-30	4	25	0	0	0	0	0
30-40	2	35	1	2	2	2	2
Total	10			-3	9	-9	21

Step I: Raw moments of d calculation

1st raw moment of d = $m_1' = \frac{1}{N}\sum fd = \frac{1}{10}\times(-3) = -0.3$

2nd raw moment of $d = m_2' = \frac{1}{N}\sum fd^2 = \frac{1}{10}\times 9 = 0.9$

3rd raw moment of $d = m_3' \frac{1}{N}\sum fd^3 = \frac{1}{10}\times(-9) = -0.9$

4th raw moment of $d = m_4' = \frac{1}{N}\sum fd^4 = \frac{1}{10}\times = 21 = 2.1$

Step II: Central moments of d calculation

$$m_2(d)=m_2^{'}-(m_1^{'})^2=0.9-(-0.3)^2=0.9-0.09=0.81$$
$$m_3(d)=m_3^{'}-3m_2^{'}.m_1^{'}+2(m_1^{'})^3$$

$$=0.9-3\times0.2\times(-0.3)+2(-.03)^3=-0.9+0.81-0.054=-0.144$$
$$m_4(d)=m_4^{'}-4m_3^{'}.m_1^{'}+6m_2^{'}.(m_1^{'})^2-3(m_1^{'})^4$$

$$=2.1-4\times(-0.9)\times(-0.3)+6\times0.9\times(-0.3)^2-3\times(-0.3)^4$$
$$=2.1-1.08+0.486-0.0243=1.482$$

Step III: Skewness and kurtosis calculation

$$\text{Kurtosis}=\frac{m_4(d)}{(m_2)_2(d)}-3=\frac{1.482}{0.81^2}-3=2.26-3=-0.74$$

$$\text{Skewness}(\gamma_1)=\sqrt{\frac{m_3^2(d)}{m_2^3(d)}}=\sqrt{\frac{(-0.144)^2}{(0.81)^3}}=-\sqrt{\frac{0.021}{0.531}}=-0.198$$

Negative sign is taken, since the sign of skewness is same as that of .

From the skewness and kurtosis value, we can conclude that the distribution is platy-kurtic and positively skewed.

5.5 Chapter Summary

In this chapter, we delve into the critical concepts of dispersion and distribution shape in statistics. We start by exploring absolute measures of dispersion, including range, variance, and standard deviation, which provide insights into the spread of data. We then transit to relative measures, emphasizing how they standardize dispersion for comparative analysis. The latter part of the chapter is dedicated to understanding the shape of distributions through moments, skewness, and kurtosis, each offering a unique perspective on the asymmetry and peakedness of data distributions. This comprehensive analysis equips readers with the fundamental tools to accurately interpret and describe data variability and distribution characteristics.

5.6 Self-Assessment

1. What do you understand by 'Dispersion'? Discuss its importance in statistics.
2. Distinguish between absolute and relative measures of dispersion.
3. Discuss Beta-coefficients and Gamma-coefficients

4. Find the range from the below data:

   i) 34, 28, 50, 68, 99, 68 ii) 2, 3, 7, 9, 10, 100

   iii) 5, 500, 200, 300, 400, 1000

5. Find Quartile Deviation and its coefficient if Q1 = 20, Q2 = 30, and Q3 = 40.

6. Calculate Quartile Deviation from the following data:

Class	10-15	15-20	20-25	25-30	30-35	35-40	40-45
Frequency	8	13	18	25	16	10	2

7. Calculate mean deviation about the arithmetic mean of the following distribution:

x	14	15	16	17	18	19	20
f	2	5	7	8	6	4	2

8. Compute Mean Deviation about median from the following data:

Class	20-29	30-39	40-49	50-59	60-69	70-79
Frequency	6	12	18	24	16	10

9. Compute standard deviation and CV from the following data:

x	20	30	40	50	60	70	Total
f	10	17	25	27	15	6	100

10. Compute skewness and kurtosis from the following distribution using moment method.

Class	0-15	15-30	30-45	45-60	60-75
Frequency	5	6	9	6	4

11. Compute skewness value based on the Pearson's first and second measures from the following data.

Class	20-29	30-39	40-49	50-59	60-69	70-79	80-89
Frequency	7	13	18	22	18	14	10

12. Create a Lorenz Curve based on the following data of the monthly income of employees of a company.

Income ('00 Rs.)	0-100	100-200	200-300	300-400	400-500
No of employee	60	80	200	140	40

Unit-VI

Correlation

6.1 Learning Objectives

After going through this chapter, you will be able to:

i) Understand the meaning, significance, and properties of correlation;

ii) Know the various types of correlation;

iii) Comprehend the computation process of various types of correlation;

iv) Appreciate the limitations of correlation.

6.2 Introduction

Correlation is a fundamental statistical concept that measures the degree of association or relationship between two or more variables. Correlation helps us to understand how changes in one variable correspond to changes in another. Correlation values range from -1 to 1; where -1 indicates a perfect negative correlation, +1 represents a perfect positive correlation and 0 suggests no linear correlation between the variables. The sign and magnitude of the correlation coefficient provide information about the strength and direction of the relationship between the variables. Correlation analysis is a multipurpose tool with various applications, such as identifying the strength of relationships between stock prices and market indices, assessing the impact of education on income levels, or determining the association between weather conditions and crop yields. Understanding correlation is essential for making informed decisions, drawing conclusions from data, and developing predictive models in numerous fields of study and practice.

6.3 Meaning of Correlation

In simple words, the correlation implies the degree of association or relationship between two or more variables. When we consider two variables at a time then it is a case of simple correlation. If we determine the degree of association between two variables taking another variable as constant then it is a case of partial correlation. When we consider more than two variables at a time then it is a situation of multiple correlation.

Correlation can be broadly classified into two types: linear and nonlinear.

Linear Correlation

Linear correlation is the most common type of correlation found in statistical analysis. It refers to a linear relationship between two variables, where the change in one variable is directly proportional to the change in the other variable. In other words, if one variable increases or decreases, the other variable also increases or decreases in a proportional manner. A linear correlation is usually represented by a straight line on a scatter plot.

For example, let's consider the relationship between the hours spent on studying and the grades obtained in a class. If the relationship is linear, we can expect that the more time a student spends on studying, the better grades they will get. This can be represented by a straight line on a scatter plot.

Linear correlation is usually measured using the Pearson correlation coefficient. The Pearson correlation coefficient ranges from -1 to +1, where -1 represents a perfect negative correlation, 0 represents no correlation, and +1 represents a perfect positive correlation. If the Pearson correlation coefficient is close to 1 or -1, it indicates a strong linear correlation between the variables.

Nonlinear Correlation

Nonlinear correlation refers to a relationship between two variables that is not linear. In other words, the relationship between the variables is not directly proportional. Nonlinear relationships can take on many different forms, including quadratic, exponential, logarithmic, and sinusoidal.

For example, consider the relationship between the amount of fertilizer used and the crop yield. Initially, an increase in the amount of fertilizer used may lead to a proportional increase in crop yield. However, after a certain level, further increases in the amount of fertilizer may lead to diminishing returns. This relationship is nonlinear and cannot be represented by a straight line on a scatter plot.

Nonlinear correlation can be measured using methods such as Spearman's rank correlation coefficient, Kendall's tau, or polynomial regression. These methods are more complex than linear correlation analysis, as they involve fitting a curve to the data rather than a straight line.

Therefore, understanding the exact type of correlation between two variables is essential in statistical analysis. Linear correlation refers to a directly proportional relationship between two variables, while nonlinear correlation refers to a relationship that is not directly proportional. Different methods are used to measure linear and nonlinear correlation, and choosing the appropriate method depends on the nature of the data and the research question at hand.

According to L.R. Connor, "If two or more quantities vary in sympathy so that movements in one tend to be accompanied by corresponding movements in others, then they are said to be correlated."

In the words of Croxton and Cowden, "When the relationship is of a quantitative nature, the appropriate statistical tool for discovering and measuring the relationship and expressing it in a brief formula is known as correlation."

According to A.M. Tuttle, "Correlation is an analysis of covariation between two or more variables."

Covariance vs. Correlation

Sometimes, we confuse two terms. These are covariance and correlation. Covariance indicates the direction of the linear relationship between variables while correlation measures both the strength and direction of the linear relationship between two variables. The correlation coefficient is the standard form of covariance.

Covariance measures the total variation of two random variables from their expected mean values. Using covariance, we can only gauge the direction of the relationship. However, it does not indicate the strength of the relationship, nor the dependency between the variables.

On the other hand, correlation measures the strength of the relationship between variables. Correlation is the scaled measure of covariance. It is dimensionless. In other words, the correlation coefficient is always a pure value and not measured in any units.

The following table depicts the differences between Covariance and Correlation:

Table 6.1 Differences between Covariance and Correlation

Basis for Comparison	**Covariance**	**Correlation**
Definition	It measures the direction of the linear association between two variables.	Correlation is a statistical measure that indicates how strongly two variables are related.
Values	The value of covariance lies in the range of $-\infty$ and $+\infty$.	The value correlation lies in the range of -1 and +1.
Change in scale	Affects covariance	Does not affect the correlation
Unit-free measure	No	Yes

6.4 Significance of Correlation

The study of correlation is of great significance because of the following reasons:

i) The correlation coefficient helps in measuring the extent of the relationship between two variables in one figure.

ii) When two variables are correlated, the value of one variable can be estimated, given the value of another. This is done with the help of regression equations.

iii) Correlation facilitates decision-making in the business world. It reduces the range of uncertainty as predictions based on correlation are likely to be more reliable.

iv) It also helps in understanding the economic behavior of the variables.

6.5 Correlation Coefficient and Causation

The degree of correlation between two or more variables can be determined using correlation. However, it does not consider the cause-and-effect relationship between variables. If two variables are correlated, it could be for any of the following reasons:

i) **Influence of the Third Variable:** The influence of a third variable can result in a high degree of correlation between the two variables. This analysis does not take into account the influence of the third variable. For example, the correlation between the yield per acre of rice and jute can be of a high degree of correlation because both are linked to the amount of rainfall. However, in reality, both these variables do not have any effect on each other.

ii) **Mutual Dependence (Cause and Effect):** It may be challenging to determine which one is the cause and which one is the effect when two variables indicate a high degree of correlation. For example, when there is an increase in the price of a commodity, it increases its demand. Here, the price is the cause, and demand is the effect. However, there is a possibility that the price of the commodity will rise due to increased demand. In that case, increased demand is the cause and the price is the effect.

iii) **Pure Chance:** It is possible that the correlation between the two variables was obtained by random chance or coincidence. This correlation is also known as spurious. For example, even if there is no relationship between the two variables (between the income of people in a society and their clothes size), one may see a strong correlation between them.

A causal relationship exists between the two occurrences, which means that causation shows that the occurrence of one event is the outcome of the other event. This concept is known as cause and effect relation.

The causation states that one occurrence results in another event. Only an experiment with a proper design can establish causation. Suppose, similar groups are given various treatments, and the outcomes of each group are examined. If any significant change in outcomes is noticed then we can say that treatment had an impact.

A correlation indicates a statistical relation between two variables. A change in one variable results in a change in another variable is causation. So, causation always implies correlation but correlation may not essentially mean causation.

6.6 Various Kinds of Correlation

The correlation will be positive or direct if an alteration in one variable's value causes an equivalent alteration in another variable. In this form of relationship, high values of one variate are linked with high values of the other, and vice versa. For example, weight and height have a positive relationship. Correlation is considered to be negative or inverse if an alteration in one variable's value causes an equivalent alteration in another variate. In this type of relation, the values of one variable are inversely related to the other. For example, the product demand and its price are negatively correlated. The variables will not be correlated if the change of values of a variable has no impact on the change of values of another variate. For example, there does not exist any relation between height and IQ.

There are numerous ways to categorize or describe correlations. The following three things are crucial:

1. Positive, Negative, and Zero correlation
2. Simple, Partial, and Multiple
3. Linear and non-linear

1. **Positive, Negative, and Zero Correlation:** The direction of the variable's change will decide whether the correlation is positive or negative.

Positive Correlation: If the values of two variables move in the same direction, that is, when one variable's value increases in regard to another variable's value and when one variable's value decreases in relation to another variable's value, the correlation is thought to be positive. For example, a positive correlation will be observed between a group of people's heights and weights, income and spending patterns, and the cost and availability of items.

Height (cm) : X	158	160	163	166	168	171	174	176
Weight (kg): Y	60	62	64	65	67	69	71	72

Negative Correlation: If the change of values of two variables differs from one another then the relationship will be called negative. If any variable increases while the other falls or if one variable declines while the other increases, the correlation is said to be negative. For example, there is a negative link between a product's price and demand.

Price of Product (Rs. Per Unit): X	6	5	4	3	2	1
Demand (In Units): Y	75	120	175	250	215	400

Zero Correlation: Zero correlation implies the absence of any relation between the variables. It indicates that a change in the value of one variable has no bearing on or altering the value of another variable. For example, the link between the weight of the person and IQ is a zero association.

2. **Simple, Partial, and Multiple Correlations:** When we only consider two variables at a time then it is a case of simple correlation. When we are studying the relation among three variables keeping the third variable as constant then it is a case of partial correlation. On the contrary, multiple correlation deals with more than two variables at a time in consideration. For example, we are interested to know the relation between wheat production per acre of land and the amount of fertilizers and rainfall utilized. In this situation, if we only consider any two variables out of three variables then we are studying simple correlation. If we consider any two variables keeping the third variable as constant then we are dealing with partial correlation. If we consider all the three variables under our consideration then we are studying multiple correlations.

3. **Linear and Non-Linear Correlation:** Whether the correlation between the variables is linear or non-linear depends on the consistency of the ratio of change between the variables. When there is a tendency for one variable's change to be proportional to another variable's change, the correlation is said to be linear.

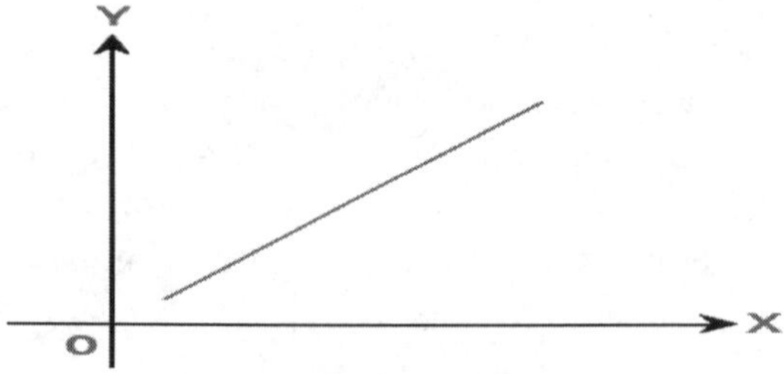

Linear Correlation

The correlation is referred to as non-linear or curvilinear when the change in one variable does not change in a way that is proportionate to the change in the other variable. For instance, the output of wheat would not necessarily double if the amount of fertilizers was doubled.

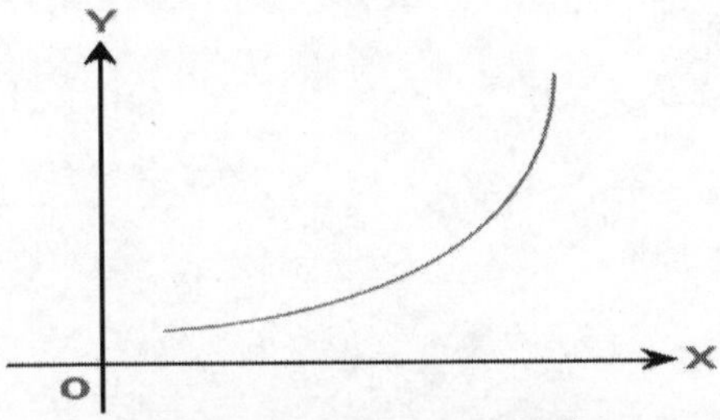

Non-Linear Correlation

In light of the movement, quantity, and the ratio of change between the variables, these are the three most significant categories of correlation. The degree to which the variables are connected must be determined using correlation methodologies, which the researcher must carefully consider. A few examples are listed below for your better comprehension.

Sl. No	Particulars	Solution
1	Price of product and its demand	Negative
2	Yield of crop and quantum of rainfall	Positive
3	No of fruits eaten and hungry a person	Negative
4	No of units produced and fixed cost per unit	Negative
5	No of girls in the class and the marks of boys	No Correlation

Within these limits, the value of correlation can be interpreted as follows:

Table 5.2 Degree of Correlation

Degree of Correlation	Positive Correlation	Negative Correlation
Perfect Correlation	+1	- 1
Very High Degree of Correlation	+.9	- .9
Fairly High Degree of Correlation	Between + 0.75 and +0.9	Between - 0.75 and -0.9
Moderate Degree of Correlation	Between + 0.25 and +0.75	Between - 0.25 and -0.75
Low Degree of Correlation	Between 0 and +0.25	Between 0 and -0.25
Zero/No Correlation	0	0

6.7 Properties of Correlation Coefficient

The main properties of correlation are appended here.

1. The value coefficient ranges between -1 and +1.

   Symbolically,-1<=r<= + 1 or | r | <1.

   Let x and y be the deviations of the X series and Y series from respective means and σ_x and σ_y be the standard deviations of the respective series.

$x_1 = X_1$ - X-bar;

$y_1 = Y_1$ – Y-bar

$\Sigma(x/\sigma_X + y/\sigma_Y)^2 = \Sigma\ [x^2/\sigma_X{}^2 + y^2/\sigma_Y{}^2 + 2xy/\sigma_X\sigma_Y]$

$=>\Sigma(x/\sigma_X + y/\sigma_Y)^2 = \Sigma\ x^2/\sigma_X{}^2 + \Sigma y^2/\sigma_Y{}^2 + 2\Sigma xy/\sigma_X\sigma_Y$ 1

Since $\Sigma x^2/n = \sigma_X{}^2$ $=>\Sigma\ x^2/\sigma_X{}^2 = n$

Similarly $\Sigma y^2/n = \sigma Y^2$ $=>\Sigma\ y^2/\sigma Y^2 = n$

Also $\Sigma xy/\sigma_X\sigma_Y = nr$

So equation 1 reduces to

$\Sigma(x/\sigma_X + y/\sigma_Y)^2 = n + n + 2nr$

$=>\Sigma(x/\sigma_X + y/\sigma_Y)^2 = 2n(1+r)$

Since Σ(x/σX + y/σY)2> = 0

=> 2n(1+r) >=0

=> 1+r >=0 => r > = - 1 2

Similarly

$\Sigma(x/\sigma_X - y/\sigma_Y)^2 = n + n - 2nr$

$=>\Sigma(x/\sigma_X - y/\sigma_Y)2 = 2n(1-r)$

Since $\Sigma(x/\sigma_X - y/\sigma_Y)2> =0$

=>2n(1-r) >= 0

=> 1- r >=0 => **1** > = **r.** 3

Combining equations 2 and 3

-1<= r <= 1

2. The origin change has no impact on correlation coefficients.

   This property shows that the correlation will not change if a constant value is deducted from all values of series X and series Y.
3. The degree of association between two variables is symmetric.
4. The correlation is not also affected by change of scale.
5. Only linear association is measured through the correlation.
6. When two variables are unrelated to one another, their correlation is zero.

Assumptions of Correlation Coefficient

Our data set must satisfy certain assumptions to get accurate results. The following assumptions are depicted here:

Assumption 1: Our data set must be measured either using a ratio or interval scale. These scales must be continuous in nature.

Assumption 2: Our two variables must have a linear relationship. There are several approaches to test the linear relationship between two variables. We advise making a scatter diagram so you can compare one variable to another and visually examine the plot to ensure linearity. Our scatter plot might resemble any of the following images depicted in Figure 5.1:

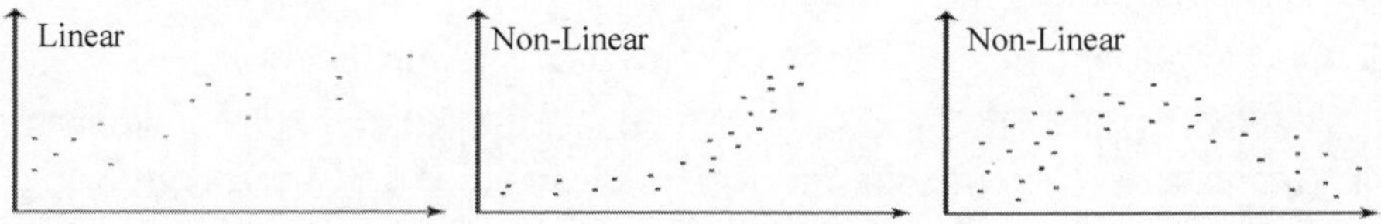

Figure 5.1 Scatter Plot

Assumption 3: There shouldn't be any notable exceptions. In the observation of 50 students' IQ tests, the mean was 100, but one student's score was 166 i.e. extremely exceptional. This is the case of an outlier. Outliers in a data set do not match the regular trend. It undoubtedly raises the mean score.

Therefore, incorporating outliers in your analysis can occasionally produce false results. Therefore, it is ideal if there are no outliers or if very few of them have outliers.

Assumption 4: The distribution of your variables should be roughly normal.

Keep in mind that the findings you obtain while computing correlation might not be valid if you do not properly evaluate these hypotheses.

6.8 Scatter Diagram

A graphical representation of bivariate data is known as a scatter diagram. With the help of a scatter diagram, we can explain various types of correlation. This diagram is a visual representation of bivariate data to find associations between two or more variables in a graphical way. A scatter diagram provides information about the type of association. In the scatter diagram, the correlation is perfect and equal to one if all points stretch down a single line. There will be a positive connection when all values go upward and a negative correlation when they move downward. The correlation is stated to be linear if all points in the diagram are on or close to a line. This technique is frequently applied in business to find out the correlation between two or more variables. On a two-dimensional graph, these two variables are displayed along the X and Y axes, and the pattern shows the connection between the two. A scatter diagram examination is the graphical representation of two variables. We should be very careful at the time of creating the graph. A scatter diagram shows how two sets of data are related to each other. Scattergram is another name for a scatter plot. A single data point is represented by a single dot in a scatter plot. A pictorial representation of the data's distribution can be observed by graphing numerous data points. Depending on how closely the dots cluster together, the data may show a clear pattern. If a straight line can be drawn from the origin to high y-values, the variables are said to be positively correlated. If a straight line can be drawn from the y-axis to the x-axis, the variables are said to be negatively correlated.

Below is a representation of the most typical and significant ones:

a) **Perfectly Positive Correlation:** In this type of relation, one unit change (increase or decrease) in one variable leads to exactly one unit change (increase or decrease) in the other variable. In this case, if we join all the coordinates of x and y then we will observe a straight line which is upward rising from the origin. This is seen in the figure below:

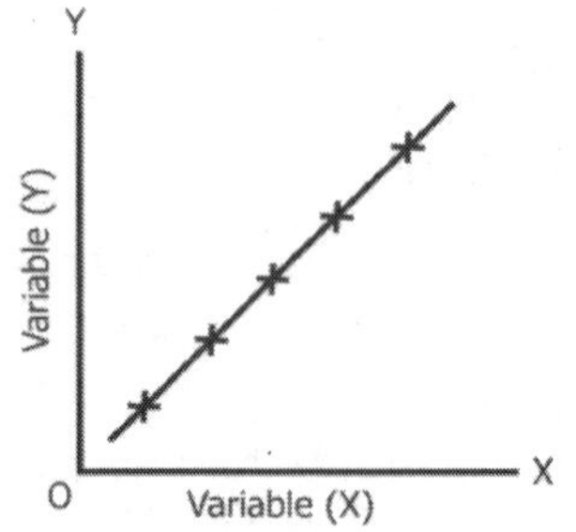

b) **Perfectly Negative Correlation:** In this type of relation, one unit increase (change) in one variable leads to exactly one unit decrease (change) in the other variable and vice-versa. In this case, if we join all the coordinates of x and y then we will observe a straight line which is downward sloping from the y axis. This is seen in the figure below:

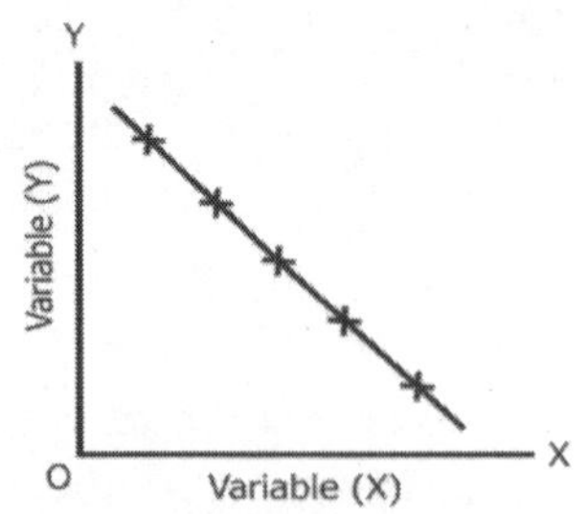

c) **High Degree of Positive Correlation:** In this type of relation, one unit change (increase or decrease) in one variable leads to not exactly one unit change (increase or decrease) of the other variable. In this case, if we join all the coordinates of x and y then we will observe a straight line which is upward rising from the origin. We will observe that all the points are very close to the imaginary straight line. This is seen in the figure below:

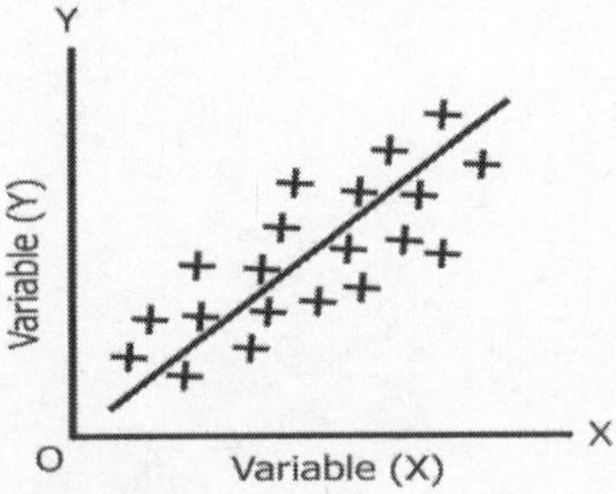

d) **High Degree of Negative Correlation:** In this type of relation, for one unit change (increase or decrease) in one variable leads to not exactly one unit change (decrease or increase) of the other variable. In this case if we join all the coordinates of x and y then we will observe a straight line which is downward slopping from the y axis. We will observe that the all the points are very close to the imaginary straight line. This is seen in the figure below:

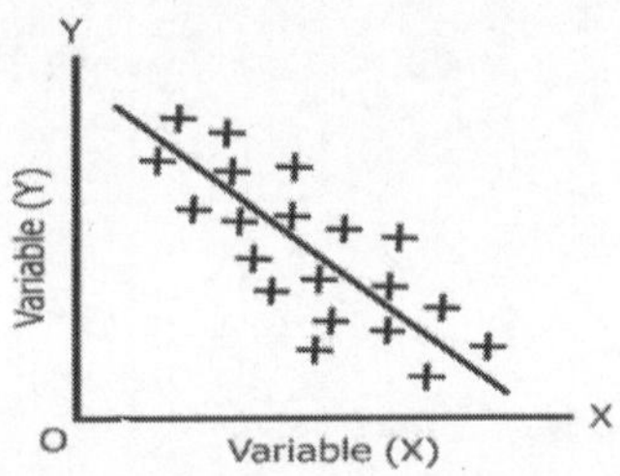

e) **Low Degree of Positive Correlation:** Points are observed to be gradually ascending from the bottom left corner to the top right corner with a modest degree of positive correlation. The illustration below demonstrates this:

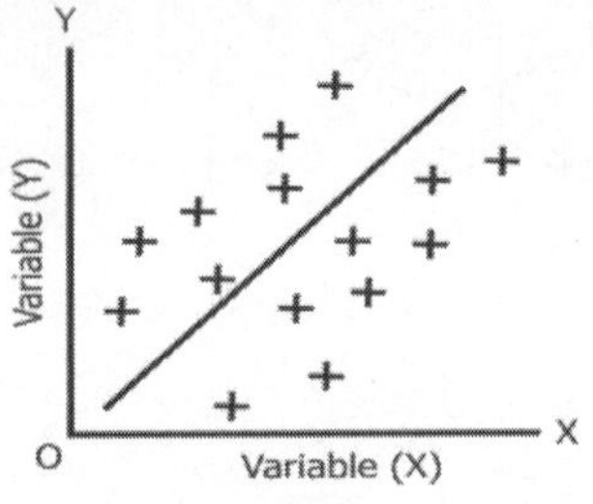

f) **Low Degree of Negative Correlation:** Points are observed to be descending steadily from the top left corner towards the bottom right corner in cases with low negative correlation. The illustration below demonstrates this:

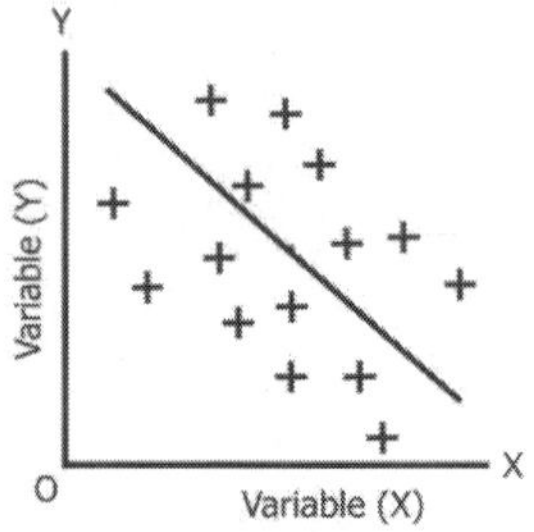

g) **No Correlation:** All of the arguments made in this case are compelled. There is no straight line formed by any of the locations. The illustration below demonstrates this:

No Correlation

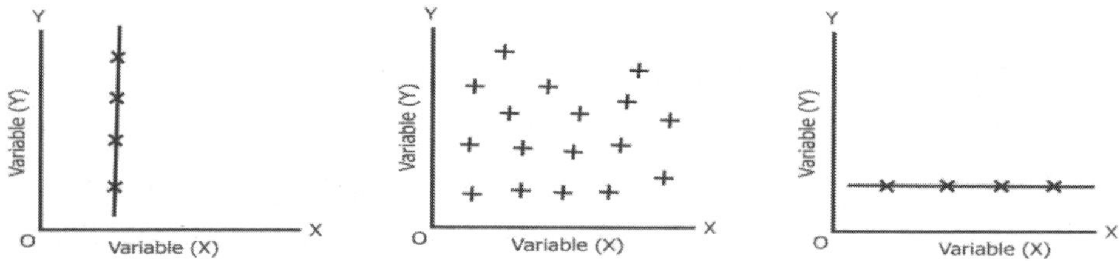

Merits of Scatter Diagram

i) It is straightforward and easy to comprehend and utilize.

ii) It is possible to study a relationship between two variables without using mathematics.

Demerits of Scatter Diagram

i) Because it is not a mathematical procedure, the outcomes are both inaccurate and non-exact.

ii) It merely provides a rough idea of the relationship.

iii) You cannot see the association for more than two variables on this chart.

Scatter Diagram Procedure

The following procedures are followed to form a scatter diagram:

i) Data pairs with a suspected association should be collected.

ii) For each pair of data, we have to provide a dot where the x- and y-axes connect.

iii) To determine whether a relationship is comprehensible, look at the pattern of the points. Because variables are associated, you may stop if the data clearly shows a line or a curve. Use regression or correlation analysis, if you'd like. If not, you must finish steps iv through vii.

iv) Divide the graph's points into four equal quadrants. If the graph contains X points:
   - Count X/2 points from top to bottom then draw a horizontal line.
   - Count X/2 points in a vertical line from left to right.
   - If there are an odd number of points, draw a line through the center.

v) Each quadrant's points should be counted. Never add up the points on a line.

vi) The quadrants that are diagonally opposed are added. Find the smaller total as well as the total of each quadrant's points.

   A = "Points in upper left" + "Points in lower right"

   B = "Points in upper right" + "Points in lower left"

   Q = "the smaller of A and B"

   N = A + B

vii) Check the trend test table for the N limit.

Table 6.2 Trend Test Table

N	limit	N	Limit
1-S	0	51-53	IS
9-11	1	54-55	19
12-14	2	56-57	20
15-16	3	5S-60	21
17-19	4	61-62	22
20-22	5	63-64	23
23-24	6	65-66	24
25-27	7	67-69	25
28-29	8	70-71	26
30-32	9	72-73	27
33-34	10	74-76	28
35-36	11	77-7S	29
37-39	12	79-80	30
4CuU	13	81-82	31
42^t3	14	83-85	32
44-46	15	86-87	33

47^8	16	88-89	34
49-50	17	90	35

- The two variables have a relationship if Q is below the limit.
- If Q exceeds or is equal to the limit, the pattern might have developed by coincidence.

Example: A manufacturing team believes that the amount of iron and the product purity are related. As seen in the scatter diagram below, purity and iron are compared to one another.

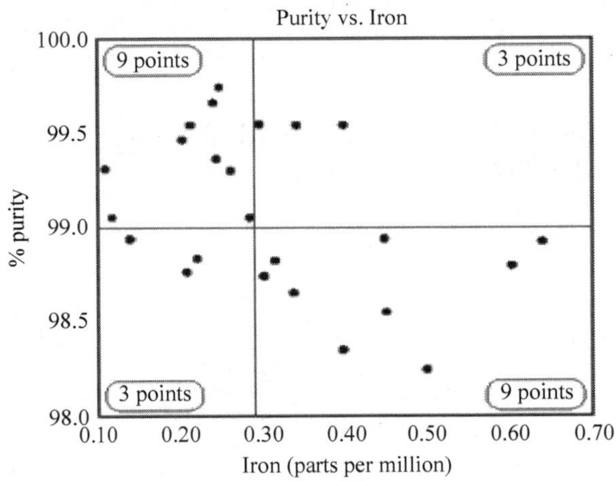

24 data points are available. Here, median lines are created so that 12 points fall on either side of them.

We must compute the following to check for a relationship:

A = "Points in upper left" + "Points in lower right" = 9 + 9 = 18

B = "Points in upper right" + "Points in lower left" = 3 + 3 = 6

Q = "the smaller of A and B" = the smaller of 18 and 6 = 6

N = A + B = 18 + 6 = 24

The limit for N will then be looked up on the trend test table. The cap is set at 6 for N = 24.Q equals the cap in this instance. Therefore, since no relationship has been shown, the pattern could have appeared by accident.

6.9 Karl Pearson's Correlation Coefficient

Karl Pearson's coefficient of correlation is denoted by r and can be used to measure correlation in the case of both individual series as well as grouped data. Pearson's correlation coefficient is a statistical indicator of the linear relationship between variables. If a straight line can be used to describe a relationship, it is said to be linear. Pearson correlation coefficient should only be used with variables that have a linear connection. It is computed using the arithmetic mean and standard deviation. A different term for the correlation is the linear correlation coefficient (r). It is a formula for figuring out how strongly two variables are linearly related. R's value ranges from -1 to 1. Use the following formula to get Pearson's correlation coefficient:

$$r_{xy} = \frac{COV_{xy}}{S_x S_y} = \frac{n\left(\sum XY\right) - \left(\sum X\right)\left(\sum Y\right)}{\left(\sqrt{n\left(\sum X^2\right) - \left(\sum X\right)^2}\right)\left(\sqrt{n\left(\sum Y^2\right) - \left(\sum Y\right)^2}\right)}$$

The correlation coefficient's value is $-1<r<1$. If the r value is outside of this range then some computational error has occurred.

We are aware that the value of r is between -1 and +1. Whether there is a positive or negative association is shown by the sign of the correlation coefficient (+,-). The size of the correlation coefficient indicates how closely the variables are related. Below is a list of the numerous 'r' value cut-offs and explanations.

Merits of Karl Pearson's Correlation Method

i) Karl Pearson's coefficient of correlation gives the exact measure of the correlation between the variables.

ii) It gives both the direction and the degree of relationship between the variables.

Demerits of Karl Pearson's Correlation Method

i) It always assumes a linear relationship between the variables.

ii) The value of the coefficient is affected by the presence of extreme values.

iii) It takes time to calculate the correlation coefficient using this method and it is a complicated method as compared to other measures of correlation.

Numerical Problems

Problem 1: Based on the information below, calculate and analyse Pearson's correlation coefficient.

Price (X)	10	12	14	15	19
Supply (Y)	40	41	48	60	50

Solution

Price: X	Supply: Y	XY	X2	Y2
11	40	440	121	1600
12	41	492	144	1681
14	48	672	196	2304
15	60	900	225	3600
19	50	950	361	2560
∑X=71	∑Y=239	∑XY=3454	∑X2=1047	∑Y2=11685

$r = \sum xy/n - (\sum x/n)(\sum y/n) / \sqrt{\sum x^2/n - (\sum x/n)^2}\sqrt{\sum y^2/n - (\sum y/n)^2}$

$r = 3454/5 - (71/5)(239/5)/\sqrt{1047/5 - (71/5)2}\sqrt{11685/5 - (239/5)^2}$

$\backslash r = 0.602$

Supply and price of the product have a positive relationship. The supply of a given good increases as its price increases.

Problem 2: The following outcomes were acquired by a computer operator who calculated the coefficient of correlation between two variates, x and y, for 25 pairs of data:

$\Sigma x = 125$, $\Sigma x^2 = 650$, $\Sigma\, y = 100$, $\Sigma y^2 = 460$, $\Sigma xy = 508$

However, it was subsequently determined that he had written down two pairings as (6, 14) and (8, 6), respectively, whereas the proper values for the pairs were (8, 12) and (6, 8). Find the accurate correlation coefficient value.

Solution: $\Sigma x = 125$

Corrected sum of $x = 125-(6+8) + (8+6)$

Corrected $\Sigma x = 125$

Corrected $\Sigma x2 = 650$

Corrected $\Sigma y = 100-(14+6) + (12+8) = 100$

Corrected $\Sigma y^2 = 460$

Corrected $\Sigma xy = 508 - (6x14+8x6) + (8x12 + 6x8)$

Corrected $\Sigma xy = 508 - 132+144 = 520$

$r = [n\sum xy - (\sum x)(\sum y)]/[\sqrt{n\sum x^2 - (\sum x)^2}]\,[\sqrt{n\sum y^2 - (\sum y)^2}]$

$r = 0.516$

Significance of Pearson's Correlation (r)

As is well known, the correlation coefficient (r) is frequently calculated to provide a sense of how closely two variables are related. When calculating the correlation coefficient (r), which is then utilized as a point estimate of the population correlation coefficient (p), we frequently take into account a sample from a sizable population. It indicates that r estimates (p). Thus, we can conclude that r is employed as a population parameter. If the premise of the normal distribution of both variables is valid, then r may be regarded as an estimate of p in this case.

The t-test is the most popular method for determining whether or not variables X and Y are associated. We use the t-test with the following hypothesis:

H0: $p = 0$

Ha: $p \neq 0$

The null hypothesis (H0) presupposes that neither of the variables X nor Y is correlated across the board. The alternative hypothesis (Ha) asserts that neither variable is substantially associated in the population, which is the opposite of the null hypothesis. Now, we can compute the t-value using the formula shown below:

$t = (r - p)/\sqrt{1 - r2/(n-2)}$

Where

r = Pearson's coefficient of correlation

p = Population correlation

n = number of observations in the sample

(n-2) = degree of freedom

Here, we have assumed population correlation (p) equals zero. We can get the value of t after putting the values of r and n. This is said to be the calculated value of t. Now, we see the table value of t the for level of significance (α) = 0.05 and the degree of

freedom = (n-2).

a) If tcal>ttable value; we will reject H0; it means r is significantly different from zero and both variables are strongly correlated in population.

b) If tcal<ttable value; we will accept H0; it means r is almost equal to zero and both variables are not correlated in population.

6.10 Spearman's Rank Correlation Coefficient

The linear correlation between the ranks given to various items based on their attributes is measured by Spearman's correlation coefficient. This approach to measuring correlation coefficients is qualitative. It is impossible to quantify qualities like beauty, honesty, ability, etc. In order to calculate the correlation coefficient, ranks are utilized.

The linear correlation between the ranks given to various items based on their attributes is measured by Spearman's correlation coefficient. This approach to measuring correlation coefficients is qualitative. It is impossible to quantify qualities like beauty, honesty, ability, etc. In order to calculate the correlation coefficient, ranks are utilised.

$R = 1 - 6\sum d^2/n^3 - n$

Here, n denotes the total number of observations, and d denotes the distinction between the ranks given to each variable.

i) **Problems where Ranks are not given:** If the ranks are missing, we must provide the data series with their own ranks. Both ascending and descending orders are acceptable for the rating. The other series must be ranked according to the same system.

ii) **Equal Ranks or Tie-in Ranks:** In such circumstances, each person should be given average ranks.

$$R = 1 - 6\,[\sum D^2 + \sum t^3 - t/12]/n^3 - n$$

Where, t = Number of individuals involved in a tie.

Interpretation of Rank Correlation Coefficient(R)

i) The rank correlation coefficient has a value between -1 and +1.

ii) If R = +1, the ranks are in perfect accord with one another and are moving in the same direction.

iii) If R is equal to -1, all of the ranks are in the same order and are moving in the opposite direction.

iv) There is no association if R is equal to 0.

Merits of Rank Correlation Coefficient

i) Calculating and comprehending it are both basic and easy.

ii) If the data are qualitative, then this approach is most appropriate.

Demerits of Rank Correlation Coefficient

i) In the case of clustered frequency distribution, this approach is inapplicable.

ii) When there are more than 30 elements, the calculations get quite time-consuming and laborious.

Numerical Problems

Problem 3: Calculate the Rank correlation coefficient using the following information:

Roll No.	1	2	3	4	5	6	7	8	9	10
Marks in Economics	78	52	48	68	50	25	90	51	45	69
Marks in Statistics	68	42	60	58	44	30	78	40	55	61

Solution

Roll No.	Marks in Economics	Marks in Statistics	Rank in Economics (x)	Rank in Statistics (y)	D = x-y	D^2
1	78	68	2	2	0	0
2	52	42	5	8	-3	9
3	48	60	8	4	4	16
4	68	58	4	5	-1	1
5	50	44	7	7	0	0
6	25	30	10	10	0	0
7	90	78	1	1	0	0
8	51	40	6	9	-3	9
9	45	55	9	6	3	9
10	69	61	3	3	0	0
						44

$R = 1 - 6\sum d^2/n^3 - n$

$= 1 - 6x44/990$

$= 1 - 264/990$

$= 1 - .27$

$= .73$

Problem 4: Calculate the Rank correlation coefficient using the following information:

Roll No.	1	2	3	4	5	6	7	8	9	10
Marks in Economics	78	52	48	68	52	25	90	52	48	69
Marks in Statistics	68	42	60	58	42	30	78	42	58	61

Solution

Roll No.	Marks in Economics	Marks in Statistics	Rank in Economics (x)	Rank in Statistics (y)	D = x-y	D2
1	78	68	2	2	0	0
2	52	42	6	8	-2	4
3	48	60	8.5	4	4.5	20.25
4	68	58	4	5.5	-1.5	2.25
5	52	42	6	8	-2	4
6	25	30	10	10	0	0
7	90	78	1	1	0	0
8	52	42	6	8	-2	4
9	48	58	8.5	5.5	3	9
10	69	61	3	3	0	0
						43.5

We see that there are a total of 4 ties between the two series, including 2 ties involving 3 players and 2 ties involving just 2 players. Hence,

$\sum t^3 - t/12 = 3^3 - 3/12 + 2^3 - 2/12 + 3^3 - 3/12 + 2^3 - 2/12 = 5$

$R = 1 - 6 [\sum D^2 + \sum t3 - t/12]/n^3 - n$

$= 1 - 6[43.5 + 5]/10^3 - 10$

$= 1 - 291/990 = .71$

Note: Ranking in the event of a tie: We see that three students (roll numbers 2, 5, and 8) had identical grades of 52 in Economics. They ought to have the same ranking in economics as a result. This tie takes place on position five. As a result, each gets assigned a rank of 6 on average [since (5 + 6 + 7)/3 = 6]. The eighth rank is thought to come next. The ranks for the other ties have also been calculated similarly.

6.11 Standard and Probable Error of Correlation

Standard Error of Correlation Coefficient (r)

We can also calculate the standard error of r using the following formula:

$SE = (1 - r^2)/\sqrt{n}$

Where, r = Correlation coefficient

n = Number of observations in the sample

Probable Error of Correlation Coefficient (r)

We can also calculate the range of degree of correlation for the entire population with the help of 'probable error'.

$P.E.r = 0.674(1-r^2)/\sqrt{n}$

Now we can estimate the lower and upper limit within which the population correlation coefficient is expected to lie. Symbolically,

$p \pm P.E.r$

6.12 Coefficient of Determination (R^2)

The most common and useful approach for determining the value of the correlation coefficient between two variables is to use the coefficient of determination. The Coefficient of Determination (r^2), which assesses the shared fluctuations in both variables, is the square of r. This measurement of the linear covariance of two variables is highly helpful.

Coefficient of Determination (r^2) = Explained Variance/ Total Variance

For example, if value of $r = 0.50$, then $r^2 = 0.25$ i.e. $r^2 = 25/ 100$

One interpretation is that only 25% of variations in the dependent variable can be explained by variations in the independent variable out of all conceivable changes (100%) The value of r^2 may vary from 0 to 1. If the value of r^2 is exactly equal to 1 then it implies that the entire fluctuation of our dependent parameter is explained by the independent parameter/s.

Regression analysis frequently uses the coefficient of determination to evaluate how well the regression model fits the data. The better the fit and the bigger the predictive value of the regression equation, the higher the value of r^2.

6.13 Partial Correlation

Many times correlation between two variables is partly due to the third variable. For example, the correlation between height and weight is due to age. In such situations, one may be interested to know the relationship between two variables ignoring the effect of third and fourth or more other variables. Partial correlation studies this type of situation. Partial correlation is the correlation between two variables after removing the linear effects of other variables on them.

Partial correlation is a statistical technique used to measure the strength and direction of a relationship between two variables while controlling for the influence of one or more additional variables. It allows researchers to assess the association between two variables while holding constant the effects of other potentially confounding variables. Partial correlation is particularly useful when you want to determine if a relationship between two variables remains significant when the influence of other variables is accounted for.

Characteristics of Partial Correlation

i) It is a study of more than two variables.

ii) One is the dependent variable and others are independent variables.

iii) We study the partial impact of one independent variable on one dependent variable keeping the other independent variable/s constant.

iv) We study the direction between positive and negative correlation.

v) We study the degree of correlation between them.

vi) Correlation ranges between 0 to 1.

Example

Let us consider the case of three variables A, B, and C. Sometimes, the relation between two variables A and B may be partly due to the relation of a 3rd variable C with both A and B. In this type of situation, one may be interested in studying the relation between A and B then the effect of C on each A and B is eliminated. This correlation is known as a partial correlation. The relation between A and B after eliminating the linear effect of C on A and B is called the partial correlation coefficient.

Formula

$$r_{123} = \frac{r_{12} - r_{13}r_{23}}{\sqrt{\left(1 - r_{13}^2\right)\left(1 - r_{23}^2\right)}}$$

Here, $r_{12.3}$ implies the partial impact of the 2nd variable on the 1st variable keeping the 3rd independent variable constant. Similarly, $r_{23.1}$ implies the partial impact of the 3rd variable on the 2nd variable keeping the 1st independent variable constant and $r_{13.2}$ implies the partial impact of the 3rd variable on the 1st variable keeping the 2_{nd} independent variable constant

Where $r_{12.3}$ implies a partial correlation between the 1st variable and the 2nd variable; r_{12} is the total correlation between the 1st variable and the 2nd variable; r_{13} is the total correlation between the 1st variable and the 3rd variable; r_{23} is the total correlation between the 2nd variable and the 3rd variable.

$$r_{23.1} = \frac{r_{23} - r_{12}r13}{\sqrt{\left(1 - r_{12}^2\right)\left(1 - r_{13}^2\right)}}$$

Where $r_{23.1}$ implies a partial correlation between the 2nd variable and the 3rd variable; r_{12} is the total correlation between the 1st variable and the 2nd variable; r_{13} is the total correlation between the 1st variable and the 3rd variable; r_{23} is the total correlation between the 2nd variable and the 3rd variable.

$$r_{13.2} = \frac{r_{23} - r_{12}r_{23}}{\sqrt{\left(1 - r_{12}^2\right)\left(1 - r_{23}^2\right)}}$$

Where $r_{13.2}$ implies a partial correlation between the 1st variable and the 3rd variable; r_{12} is the total correlation between the 1st variable and the 2nd variable; r_{13} is the total correlation between the 1st variable and the 3rd variable; r^{23} is the total correlation between the 2nd variable and the 3rd variable.

Interpret the Result: The resulting partial correlation coefficient, $r_{12.3}$ quantifies the strength and direction of the relationship between the 1st variable and the 2nd variable after adjusting for the influence of the 3rd variable. A positive value indicates a positive relationship, while a negative value indicates a negative relationship. The magnitude of the coefficient indicates the strength of the relationship.

For example, if the partial correlation coefficient between exercise hours and weight loss while controlling for age is strongly positive (e.g., 0.70), it suggests that exercise is significantly associated with weight loss, even after considering the potential influence of age. If the coefficient is close to zero, it indicates that the relationship is weak or non-existent when age is taken into account.

Partial correlation allows researchers to better understand the specific relationship between two variables while accounting for the impact of other relevant factors, providing more robust and nuanced insights into the relationships within their data.

Given r_{12}= 0.87, r_{13} = 0.66, and r_{23} = 0.57, compute the correlation between X_1 and X_2 while controlling for X_3. N=266 for all r.

We are looking for the relationship between two variables while controlling for one other, so this is a first-order partial correlation.

$$r_{12.3} = \frac{r_{12} - r_{13}r_{23}}{\sqrt{\left(1 - r_{13}^2\right)\left(1 - r_{23}^2\right)}}$$

$$r_{12.3} = \frac{0.87 - 0.66(0.57)}{\sqrt{(1 - 0.66^2)(1 - 0.57^2)}} = \frac{0.4938}{\sqrt{0.3810}} = 0.80$$

The correlation between X_1 and X_2 while controlling for X_3 is 0.80

This is a strong positive correlation. However, you can see that the correlation between X_1 and X_2 dropped from the 0.87 to 0.80 after the third variable was taken into account.

Given r_{12}= -0.70, r_{13}= 0.60, and r_{23}= -0.52, compute the correlation between X1 and X_2 while controlling for X_3. N=40 for all r.

We are looking for the relationship between two variables while controlling for one other, so this is a first-order partial correlation.

$$r_{12.3} = \frac{r_{12} - r_{13}r_{23}}{\sqrt{(1 - r_{13}^2)(1 - r_{23}^2)}}$$

$$r_{12.3} = \frac{-0.70 - (0.60)(-0.52)}{\sqrt{(1 - 0.60^2)(1 - 0.52^2)}} = \frac{-0.388}{\sqrt{0.4669}} = -0.5678$$

The is a moderate negative correlation. Again, the magnitude of the relationship between X1, and X2 decreased from the -0.7 to -0.5678, after the third variable was controlled for.

6.14 Multiple Correlation

The distinction between simple, partial, and multiple correlations is based on the number of variables studied. When only two variables are studied, it is a case of simple correlation. For example, when one studies the relationship between the marks secured by students and the attendance of students in class, it is a problem of simple correlation.

In the case of partial correlation one studies three or more variables but considers only two variables to be influencing each other and the effect of other influencing variables being held constant. For example, in the above example of the relationship between student marks and attendance, the other variables influencing such as effective teaching of teacher, use of teaching aids like computers, smart boards, etc. are assumed to be constant.

Multiple correlation is different from simple and partial correlation. When three or more variables are studied, it is a case of multiple correlation. For example, in the above example if the study covers the relationship between student marks, attendance of students, the effectiveness of the teacher, use of teaching aids, etc. it is a case of multiple correlations.

In many studies, data are available on more than two variables, and the value of one variable seems to be influenced by two or more variables. Multiple correlation is the study of the combined influence of two or more variables on a single variable. Let us take another example:

The production of a crop may depend upon the amount of rainfall, quality of seeds, quantity of fertilizers used, method of irrigation, etc. Similarly, the performance of students in university exams may depend upon his/her IQ, mother's qualification, father's qualification, parents' income, number of hours of studies, etc. Whenever we are interested in studying the joint effect of two or more variables on a single variable, multiple correlation gives the solution to our problem. In fact, multiple correlation is the study of the combined influence of two or more variables on a single variable.

Characteristics of Multiple Correlation

i) It is a study of more than two variables. One is assumed to be a dependent variable and others are independent variables.

ii) A multiple correlation coefficient is a non-negative coefficient. The value of multiple correlation ranges between 0 to 1.

iii) We study the multiple impact of independent variables on dependent variable.

iv) We study the degree and direction of correlation among the variables.

v) Multiple correlation coefficient is always greater or equal to any total correlation coefficient. If R1.23 implies a multiple correlation coefficient the value of $R_{1.23} \geq r_{12}$ or r_{13} or r_{23}.

vi) If the multiple correlation coefficient is 1, then the association is perfect, and the multiple regression equation may said to be a perfect prediction formula.

vii) If the multiple correlation coefficient is 0, the dependent variable is uncorrelated with other independent variables. From this, it can be concluded that the multiple regression equation fails to predict the value of the dependent variable when the values of independent variables are known.

Formula

i) When we study the multiple impacts of 2nd and 3rd variables on the 1st variable then the following formula is used:

$$R_{1.23} = \sqrt{\frac{r_{12}^2 + r_{13}^2 - 2r_{12}r_{13}r_{23}}{1 - r_{23}^2}}$$

Where $R_{1.23}$ = Multiple correlation coefficient; r_{12} is the total correlation coefficient between 1st and 2nd variable; r_{13} is the total correlation coefficient between 1st and 3rd variable; and r_{23} is the total correlation coefficient between 2nd and 3rd variable.

ii) When we study the multiple impacts of 1st and 3rd variables on the 2nd variable then the following formula is used:

$$R_{2.13} = \sqrt{\frac{r_{12}^2 + r_{23}^2 - 2r_{12}r_{13}r_{23}}{1 - r_{13}^2}}$$

Where $R_{2.13}$ = Multiple correlation coefficient; r_{12} is the total correlation coefficient between 1st and 2nd variable; r_{13} is the total correlation coefficient between 1st and 3rd variable; and r_{23} is the total correlation coefficient between 2nd and 3rd variable.

iii) When we study the multiple impacts of 1st and 2nd variables on 3rd variable then the following formula is used:

$$R_{3.12} = \sqrt{\frac{r_{13}^2 + r_{23}^2 - 2r_{12}r_{13}r_{23}}{1 - r_{12}^2}}$$

Where $R_{3.12}$ = Multiple correlation coefficient; r_{12} is the total correlation coefficient between 1st and 2nd variable; r_{13} is the total correlation coefficient between 1st and 3rd variable; and r_{23} is the total correlation coefficient between 2nd and 3rd variable.

Example 2: From the following data, obtain $R_{1.23}$, $R_{2.13}$, and $R_{3.12}$

X_1	2	5	7	11
X_2	3	6	10	12
X_3	1	3	6	10

Solution

To obtain multiple correlation coefficients R1.23, R2.13, and R3.12, we use the above formulae:

We need r_{12}, r_{13}, and r_{23} which are obtained from the following table:

Sl. No	X^1	X^2	X^3	$(X1)^2$	$(X2)^2$	$(X3)^2$	$X1X^2$	$X1X^3$	$X2X^3$
1	2	3	1	4	9	1	6	2	3
2	5	6	3	25	36	9	30	15	18
3	7	10	6	49	100	36	70	42	60
4	11	12	10	121	144	100	132	110	120
Total	25	31	20	199	289	146	238	146	201

Now, we get the total correlation coefficient r12, r13, and r23

$$r_{12} = \frac{N(\sum X_1 X_2)-(\sum X_1)(\sum X_1)(\sum X_2)}{\sqrt{\left\{N(\sum X_1^2)-(\sum X_1)^2\right\}\left\{N(\sum X_2^2)-(\sum X_2)^2\right\}}}$$

$$r_{12} = \frac{(4\times 238)-(25)\times(31)}{\sqrt{\{(4\times 199)-(25)\times(25)\}\{(4\times 289)-(31)\times(31)\}}}$$

$$r_{12} = \frac{177}{\sqrt{\{171\}\{195\}}} = \frac{177}{182.61} = 0.97$$

$$r_{13} \frac{N(\sum X_1 X_3)-(\sum X_1)(\sum X_1)(\sum X_3)}{\sqrt{\left\{N(\sum X_1^2)-(\sum X_1)^2\right\}\left\{N(\sum X_3^2)-(\sum X_3)^2\right\}}}$$

$$r_{13} \frac{(4\times 169)-(25)\times(20)}{\sqrt{\{(4\times 199)-(25)\times(25)\}\{(4\times 146)-(20)\times(20)\}}}$$

$$r_{13} = \frac{176}{\sqrt{\{171\}\{184\}}} = \frac{176}{177.38} = 0.99$$

$$r_{23} = \frac{N(\sum X_2 X_3)-(\sum X_2)(\sum X_3)}{\sqrt{\left\{N(\sum X_2^3)-(\sum X_2)^2\right\}\left\{N(\sum X_3^2)-(\sum X_3)^2\right\}}}$$

$$r_{23} \frac{(4\times 201)-(31)\times(20)}{\sqrt{\{(4\times 289)-(31)\times(31)\}\{(4\times 146)-(20)\times(20)\}}}$$

$$r_{23} = \frac{184}{\sqrt{\{195\}\{184\}}} = \frac{184}{189.42} = 0.97$$

Now, we calculate $R_{1.23}$, $R_{2.13}$, $R_{3.12}$

We have, $r_{12} = 0.97$, $r_{13} = 0.99$, and $r_{23} = 0.97$, then

$$R_{1.23} = \sqrt{\frac{r_{12}^2 + r_{13}^2 - 2r_{12}r_{13}r_{23}}{1 - r_{23}^2}}$$

$$R_{1.23} = \sqrt{\frac{0.97^2 + 0.99^2 - 2 \times 0.97 \times 0.99 \times 0.97}{1 - 0.97^2}}$$

$$R_{1.23} = \sqrt{\frac{0.058}{0.059}} = \sqrt{0.98} = 0.99$$

$$R_{2.13} = \sqrt{\frac{r_{12}^2 + r_{23}^2 - 2r_{12}r_{13}r_{23}}{1 - r_{13}^2}}$$

$$R_{2.13} = \sqrt{\frac{0.97^2 + 0.97^2 - 2 \times 0.97 \times 0.99 \times 0.97}{1 - 0.99^2}}$$

$$R_{2.13} = \sqrt{\frac{0.19}{0.20}} = \sqrt{0.95} = 0.97$$

$$R_{3.12} = \sqrt{\frac{r_{13}^2 + r_{23}^2 - 2r_{12}r_{13}r_{23}}{1 - r_{12}^2}}$$

$$R_{3.12} = \sqrt{\frac{0.99^2 + 0.97^2 - 2 \times 0.97 \times 0.99 \times 0.97}{1 - 0.97^2}}$$

$$R_{3.12} = \sqrt{\frac{0.58}{0.591}} = \sqrt{0.981} = 0.99$$

6.15 Limitations of Correlation

It must be applied with extreme caution to prevent erroneous interpretations or conclusions. The following are the most frequent errors that we commit:

i) By examining the correlation coefficient, we may get a broad grasp of the linear connection between the variables (r). A rise in the value of r from 0 to 1 indicates a strengthening of the linear connection between the variables. A value of r = 0 does not indicate that a relationship does not exist. Both variables in this situation are not linearly related to one another, but they may still be connected in other ways.

ii) For the values of r = 0.8 and r = 0.4, we cannot say that the degree of connection between the variables in the first example is simply twice as large as it is in the second. It implies that the correlation coefficient does not adhere to the proportionality principle.

iii) R = 0.6 does not necessarily suggest that correlation explains 60% of the total variation. Instead, r = 0.6 indicates that correlation only partially accounts for both variables' variability (r2 = 0.36).

iv) One can perceive a substantial correlation between the variables as a cause-and-effect connection (causation). In reality, correlation does not speak to cause and effect. Just simple regression analysis can be used to establish causation when there are only two variables.

v) We commit a very typical mistake in the interpretation of the correlation coefficient when we infer that there is a significant link between the variables when there isn't one. For instance, the number of accidents in Mumbai and the sales of Hero bikes in New Delhi. Due to their comparable movements, both variables appear to be connected, however, this is not feasible.

6.16 Summary

The correlation is referred to as linear when there is a propensity for one variable's change to be proportional to another variable's change. Correlation analysis is used to determine whether there is a relationship between the variables under investigation. Additionally, a correlation coefficient is employed to give a signed numerical number indicating the strength and direction of the link. Correlation assesses the similarity between two sets of measurements or results. Correlation simply shows how one variable relates to another.

Correlation is basically applied to determine the degree of association between 2 variables. A measurement of the link between 2 variables and the amount to which changes in any variable have an impact on other variable's changes is called the correlation coefficient. The value coefficient ranges between -1 and +1.

A correlation indicates a statistical relation between 2 variables. A change in one variable result in a change in another variable is causation. So, causation always implies correlation but correlation may not essentially mean causation.

Scatter diagrams are helpful in figuring out how two variables are related to one another. This connection might exist between two causes or between an effect and a cause. It may correlate positively, negatively, or not at all. While the second variable depends on the first, the first variable is independent. You alter the independent variable and keep an eye on the dependent variable's changes to assess the relationship's pattern. Two independent variables can be present in a scatter diagram.

In correlation, the direction and strength of correlations between variables are examined and quantified. A scatter diagram provides a visual representation of a connection that is not limited to linear connections. Both rank correlation and Pearson's correlation evaluate the linearity of the relationship between the variables. Rank correlation can be utilized when the variables can't be measured exactly.

Correlation information may be used to forecast the degree and direction of change in a variable when the correlated variable changes. Some of the most common forms of correlations are positive, negative, zero, simple, multiple, partial, linear, and non-linear correlations.

Partial correlation is a statistical technique used to measure the strength and direction of a relationship between two variables while controlling for the influence of one or more additional variables. Multiple correlation is the study of the combined influence of two or more variables on a single variable.

6.17 Self-Assessment

1. What do you mean by correlation?
2. What is the significance of correlation?
3. State the important properties of correlation.
4. How causation is different from correlation?
5. Discuss various types of correlation with suitable examples.
6. Explain the scatter diagram.
7. Discuss different types of scatter diagrams.
8. State the important properties of the correlation coefficient.
9. How does rank correlation differ from product-moment correlation?
10. Discuss the drawbacks of correlation analysis.
11. Distinguish between partial correlation and multiple correlation with suitable examples.
12. State the important properties of partial correlation.
13. Mention the important properties of multiple correlation.
14. Calculate the linear correlation coefficient for the following data. X = 4, 8, 12, and 16, and Y = 5, 10, 15, and 20.

15. Marks obtained by 5 students in algebra and trigonometry as given below:

Algebra	15	16	12	10	8
Trigonometry	18	11	10	20	17

16. In bivariate distribution, r12 = .6, r23 = r31 = 0.54, then calculate R1.23.
17. If $r_{12} = .6$, $r_{23} = r_{31} = 0.54$, then calculate the value of $r_{12.3}$.

Unit-VII

Regression

7.1 Learning Objectives

After comprehending this unit, we will be capable to:

i) Understand the conception of regression;
ii) Appreciate the significance of regression;
iii) Understand the difference between regression and correlation;
iv) Know about the regression lines;
v) Anticipate the difference between simple regression and multiple regression.

7.2 Introduction

Correlation is followed by linear regression. Regression is applied for forecasting the value of an unknown variable with the help of a known variable. The forecasted variable is called the dependent variable. On the contrary, an independent variable is used for forecasting the value of a dependent variable. If you desire to know, for instance, if exam achievement can be anticipated based on review time, whether cigarette consumption can be predicted based on smoking length, and so on, you may use linear regression. You must utilize multiple regressions if we deal with more than one independent variable rather than simply one.

Suppose, you desire to predict sales for your company and you conclude that your company's sales go up and down depending on changes in GDP. In this case, you will consider sales to be the dependent parameter and changes in GDP to be the independent parameter.

You may develop a linear equation to forecast the sales of your company with respect to GDP in the following way:

$Y = a + b*X$

Where,

Y = dependent variable;

a = Y-intercept, i.e., intercept along the y-axis;

b = regression coefficient of Y on X; X = independent variable.

The intercept, or 'a', is the y value (dependent variable) if the x value (independent variable) is zero. So, if there was no change in variable GDP, your company would still make some sales. The value of 'b' indicates that if there is a 1% increase in GDP, the sale is possibly to go up (or down) by 'b' units. If there is a positive coefficient then it will go up or else go down.

7.3 Meaning of Regression

A statistical method for determining the relations between dependent parameters and one or more independent parameters is called regression analysis. For forecasting purposes, regression is frequently utilized. Regression is used in many different disciplines, including management, business, engineering, and social science.

Take an example to comprehend the concept of regression. Presume you are holding the post of marketing manager for Disney Toys and you have collected the data concerning the advertising expenditure and the sales figures. Now, you desire to know if there is any relation between advertising expenditure and sales. Whatever the money we are spending for advertising purposes is a sort of controlled variable. This expenditure is under your control only. As a marketing manager, you have the controlling power to determine the advertising cost of your company. But, the sales quantity is not a controlled variable. You cannot control the sales figures. So, the sale is a dependent parameter. It relies on the advertisement cost i.e. the quantum of money spent on advertisement.

So, in the above-stated case, the sale is a dependent parameter and the advertising cost is an independent parameter.

Regression is referred to be simple linear regression if there is only one independent variable and there is a linear trend between the dependent and independent variables. On the contrary, if we consider more than one independent variable in the study then it is a case of multiple regression. Suppose the production of a crop depends on multiple factors like rainfall, temperature, soil quality, fertilizer, pesticides, etc. Sometimes we may have categorical dependent parameters. If the nature of the dependent parameter is categorical especially binary character then we will go for logistic regression.

Let us talk now about scatter plots to comprehend the matter in a very lucid way taking an example. Suppose, we have five observations relating to the advertising expenditure and the sales figure of a company. The figures are provided below:

Advertising Cost in Thousands (X)	Sales in Quantity (Y)
10	10
20	10
30	20
40	20
50	40

Here, advertising cost will be considered as an independent parameter which will be represented as X and the sales figure will be treated as a dependent parameter will be denoted as Y. We will get a scatter plot by plotting the above-mentioned data on the X Y plane. A scatter plot implies a mathematical diagram to show the values of two variables. We have already learned how to sketch scatter plots in the previous unit.

If the scatter plot indicates a sort of linear relation between the variables then we will definitely go for the linear model. So, if we find a particular line trend in the data set then it gives us assurance that the linear model may be suitably fitted here. But, if the scatter plot indicates a sort of nonlinear relation between variables X and Y then we should go for like a quadratic fit, a cubic fit, or the higher loaded polynomial fit.

In this unit, we will only discuss simple linear regression with a single regressor X that has a linear relationship with a response Y. In this model, variable X i.e. regressor X must have linear association with response variable i.e. Y. The form of this model is as follows:

$Y = \beta o + \beta_1 X + €$

In the above equation, Y indicates the response variable, X indicates the regressor variable, βo implies the intercept, β_1 indicates the slope of the regression line, and € implies the random error term. In this model, Y is equal to beta naught plus beta 1 X plus epsilon. Here, Y is the response variable, X is the regressor variable, beta naught is called intercept, beta 1 is called the slope of regressor variable X, and epsilon is a random error component. βo implies the average value of response when the independent variable takes zero value. β_1 implies the rate of change in the average value of response when there is a unit change in the value of X.

If we write the model in the form of only $Y = \beta o + \beta_1 X$ then it implies the exact relation between Y and X. In reality, all the coordinates of Y and X do not lie exactly on the exact trend equation. The coordinate of each pair of observations might be slightly upper or lower on the exact trend line. If the value of any coordinate lies on the upper portion of the trend line then the difference between the observed value and the expected value will be positive and the value of the random error term will be positive. On the contrary, if the value of any coordinates lies on the lower portion of the trend line then the difference between the observed value and expected value will be negative and the value of the random error term will be negative. Therefore, we can say that the value of the error term may be positive or negative. So, in a regression equation, the error term is considered. If we consider the error term in our regression model then it can be represented as $Yi = \beta o + \beta_1 Xi + €i$. Epsilon i is a random variable with zero mean and variance sigma square, which is unknown.

For simplicity purposes, we will consider the following form to represent the simple linear regression model i.e. $Y = \beta o + \beta_1 X$.

Here, βo and β_1 are the unknown parameters of the regression equation. These values are constant. If we can compute the value of these unknown parameters then we can easily determine the value of our response variable for a given value of an independent variable.

Another interesting point is that out of two variables Y and X, we can decide the value of X as per our decision. Therefore, X is not a random variable. On the contrary, the value of Y is a dependent parameter and it may take any value which is dependent on the value of X. So, Y is basically a random variable.

Let us discuss now the simple linear regression model where Y is equal to beta naught plus beta 1 X plus epsilon. It means that for a given value of X say advertising cost, the corresponding value of Y say sales quantity will consist of the value beta naught plus beta 1 X plus an amount of epsilon.

Now, we will make some basic assumptions of a simple linear model. You need to understand the different assumptions that your data must meet in order for linear regression to give you a valid result. First, let's take a look at these assumptions. We discuss these assumptions here.

Assumption 1: If we have been given a set of observations of Xi Yi where i = 1,2,3,….n then the epsilon i i.e. the error term will be a random variable with zero mean and variance sigma square. The expectation of epsilon i is equal to 0 and the variance of epsilon i is equal to sigma square.

Assumption 2: Epsilon i is a normally distributed random variable with a mean 0 and variance sigma square.

Assumption 3: You should measure your two variables using a continuous scale of measurement.

Assumption 4: The two variables must be related in a linear fashion. There are several approaches to determine if your two variables have a linear connection. We advise making a scatter plot so you may compare one variable to the other, and then visually examine the scatter plot to ensure linearity. Your scatter plot might resemble one of the examples below:

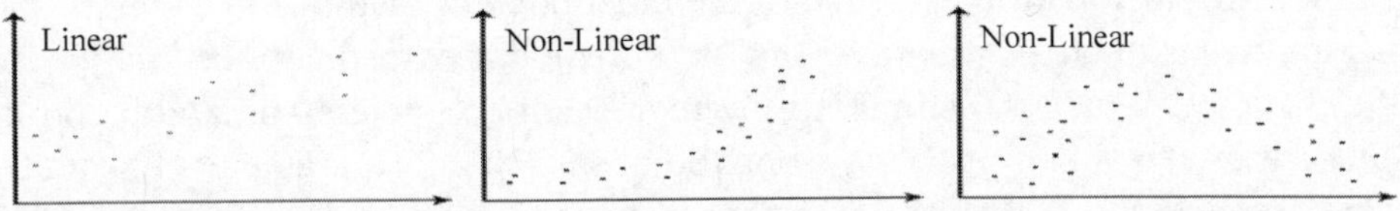

Assumption 5: There shouldn't be any notable exceptions. An observed data point that deviates significantly from the value predicted by the regression equation for the dependent variable is referred to as an outlier. An outlier will therefore be a point on a scatter plot that is far from the regression line and shows a big residual, as shown below:

Outlier in Scatter Plot

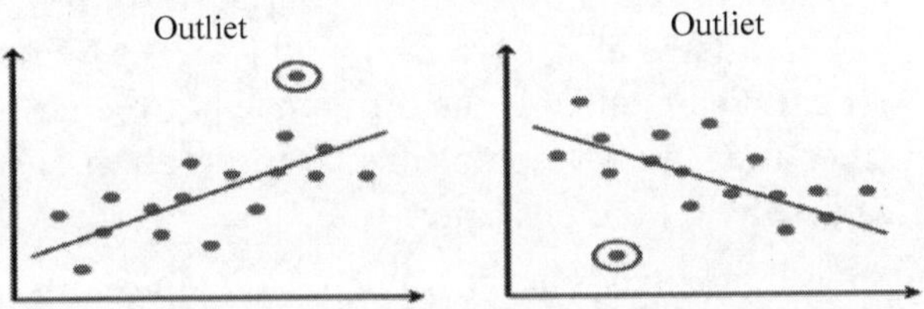

The problem with outliers is that they can have a negative effect on the regression analysis (e.g., reduce the fit of the regression equation) that is used to predict the value of the dependent (outcome) variable based on the independent (predictor) variable.

Assumption 6: You should have independence of observations, which you can easily check using the Durbin-Watson statistic.

Assumption 7: The variations along the line of best fit must exhibit homoscedasticity, which is the property where the variances are similar along the line as you move along the line. Look at the three scatter plots below for three straightforward examples: two of data that contradict the assumption (Heteroscedasticity), and one of data that upholds it (called homoscedasticity).

Scatter Plots on Heteroscedasticity and Homoscedasticity

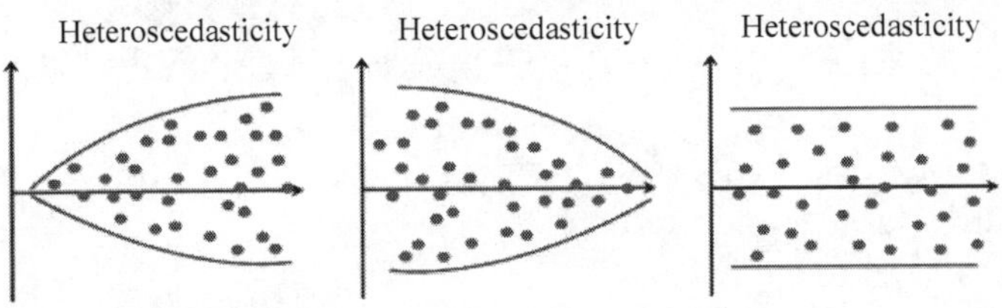

Remember that if you do not test these assumptions correctly, the results you get when running a Regression equation might not be valid.

Estimation of Parameters: One of the most popular methods to estimate the parameters is the least squares estimation method. Least squares estimation of the parameters means, estimating the regression coefficients βo and β_1. βo is called intercept and β_1 is called slope. So, it says that the parameters βo and β_1 are unknown parameters and must be estimated using the given data set of Y and X.

So, using a set of observations we have to determine the best fitted line. Suppose the fitted model is Y hat, which is equal to βo hat plus β_1 hat x X. The line fitted by the least square is the one that makes the sum of squares of all vertical discrepancies as small as possible. So, this is the main idea behind the least square fit method.

The vertical discrepancy is a residual for each observation. Suppose we have five observations (X1, Y1), (X2, Y2), (X3, Y3), (X4, Y4), and (X5, Y5). For example, for 1st observation vertical discrepancy is the difference between Y1 and Y1 hat, and so on. Y1 implies the actual value of the 1st observation and Y1 hat implies the estimated value of Y of the best fitted line.

The least-square technique fits a model such that the value of e i square for i equal to 1 to n becomes minimum. This e i square value is called the residual sum of square or SS residual. For estimating the unknown parameters βo hat and β_1 hat, we will take the help of two normal equations.

We can summarize the above discussion in a very simple way. Let (x_1, y_1), (x_2, y_2), ...(x_n, y_n) be n pairs of observations of two variables x and y. We want to establish a linear relation between x and y in the form y = a + bx where a and b are two constants that are determined by the method of least squares. For estimating the unknown parameters a and b, we will consider the following normal equations:

$\sum y = na + b\sum x$ and $\sum xy = a\sum x + b\sum x^2$

If we solve the two normal equations then we can easily get the value of two unknown parameters a and b.

Advantages of Regression

i) Regression analysis helps in making predictions and forecasting for business in the near and long term.

ii) It supports business decisions by providing necessary information related to dependent targets and predictors.

iii) Regression analysis enables businesses to correct errors by doing proper analysis of results derived from decisions.

iv) Businesses use regression analysis techniques for identifying new opportunities in the market.

7.4 Significance of Regression

Regression analysis aids an organization in better decision-making by enabling them to comprehend what their data points signify and make use of them accordingly. You will learn through this analysis how changing one of the independent variables while maintaining the other independent variables' values affects the typical value of the dependent variable. This potent statistical tool is used by data specialists and business analysts to eliminate irrelevant variables and choose the crucial ones. Regression analysis can be used in many different ways by an organization, some of which are covered here.

i) **Decision-Making:** Organizations, in order to run smoothly as well as efficiently, need better decisions and must understand the effects of the decisions taken. Organizations collect data about sales, investments, expenditures, and other parameters and analyze it for improvement. The regression analysis helps the organizations to make sense of the data which is then used for gaining insights into an organisation. Business analysts and data professionals use regression analysis to make strategic business decisions.

ii) **Optimization of Business:** Regression analysis is used to transform the gathered data into useful knowledge. Organizations are switching to data-driven decision-making, which does away with antiquated methods like guesswork or assuming a hypothesis and ultimately enhances organizational productivity. This examination offers the organization's management unit useful advice. Given the abundance of data, it is possible to analyze and comprehend the data in order to get useful insights and operate shrewdly.

iii) **Predictive Analysis:** Organizations use regression analysis in order to predict future events. Multivariate linear regression is used for various important purposes such as forecasting sale volumes, creating growth plans, etc.

iv) **Risk Analysis:** Organizations utilize regression analysis to evaluate risk in the financial sector and other similar sectors and serve as a roadmap for making important business choices.

v) **Understand the Failures:** In addition to helping an organization analyze data, regression analysis also enables it to comprehend failures and draw lessons from them for future analysis. The decision-making process is essentially supported quantitatively by the regression analysis. Predicting success can be said as one of the main assets of regression analysis such as analyzing the data points of previous sales data as well as current sales data in an organization to understand and predict future success.

7.5 Regression vs. Correlation

The key differences between these two aspects are given below.

Correlation vs. Regression

Basis of Difference	Correlation	Regression
Meaning	A statistical measure known as correlation identifies the association or co-relationship between two variables.	Regression shows the mathematical relationship between an independent variable and any dependent variable.
Utility	Utilized to illustrate the linear relationship between two variables.	It is used to find the optimal line to fit and to calculate a variable's value based on how it relates to another.
Dependent / Independent variables	The two are identical to one another. Both variables are reliant on one another.	Regression analysis makes use of the differences between the two variables. While the other variable is dependent, one is independent.
Indicator of	It demonstrates the degree and mode of interaction between two variables.	Regression shows how the value of the estimated variable (y) is affected by every unit change in the value of the known variable (x) (y).
Objective	To ascertain the numerical representation of the link between the variables.	Based on the values displayed by fixed variables, estimate the values of random variables.
Purpose	The main goal is to forecast the most trustworthy forecasts.	The main goal is to use the known variable to forecast or estimate the value of any unknown variable.
Scope	There are a few uses for correlation analysis.	A wider range of applications is provided by regression analysis.
Range	The coefficients can be between -1.00 and +1.00.	If byx > 1, then bxy < 1 in regression analysis.
Responding Nature	Any shift in Origin or change in Scale is irrelevant to the correlation coefficient.	Although the shift in origin has no effect on the regression coefficient, the scale change does.
Nature of Coefficient	The coefficient of correlation is reciprocal and symmetrical.	Asymmetricity of the regression coefficient is absent.
Exceptional Cases	Some correlation analysis may include nonsensical correlation.	Regression analysis does not use nonsensical regression.

Basis of Difference	Correlation	Regression
Mathematical treatment.	Not particularly helpful for complex mathematical procedures.	For treating advanced mathematical problems, it is frequently utilized.
Measures	The degree or extent to which any two variables move together is measured in this form of analysis.	It shows the nature of the current linear relationships between two variables as well as their fundamental level. Regression is the description of one variable as a linear function of another one.
Relationship	It is limited to only those variables that already have linear relationships. Correlation does not show how the variables' causes and effects interact.	Both linear and non-linear relationships are included in it. A functional link is established, and it is shown that there is a cause-and-effect relationship between the two.
Variables	X and Y are random variables, as well.	In regression, x is a random variable while y is a fixed variable. At times, both variables may be like random variables.
Coefficient	The correlation coefficient is used as a comparative metric.	Typically, the regression coefficient is an absolute number.

Relation Between Correlation Coefficient and Regression Coefficient

We know that the regression coefficient of x on y = bxy = r. σy / σx and regression coefficient of y on x = byx = r. σx / σy

\ bxy.byx = r. (σy / σx) x r.(σx / σy) = r^2

\ r = √bxy.byx (Here r is the correlation coefficient)

The formula for correlation coefficient when two regression coefficients are given; r = √bxy*byx

The coefficient of correlation is the geometric mean (i.e. square root of product) of regression coefficients. Hence when regression coefficients are 0.2 and 1.8 then correlation coefficient =√.2*1.8 = 0.4.

7.6 Regression Lines

The goal of regression is to estimate the optimal value of one variable for a specific value of the other variable using a set of predetermined pairs of observations. We require a relationship between the two variables for this objective. A regression equation or estimating equation is the name given to this relationship. Here, we will just take into account the most straightforward type of relationship, a linear relationship. Thus, the equation for the regression will be known as the linear regression equation.

There are two different kinds of regression equations: (1) y on x regression equations and (2) x on y regression equations. Regression lines or estimating lines, which are produced using the least squares method, are the geometric representations of the linear regression equations.

1. **Regression Line of y on x:** The equation will be written as y = a + bx, where a and b are two constants that are found using two standard equations. We may readily determine the values of these two constants if we solve the two standard equations.

$\sum y = na + b \sum x$

$\sum xy = a \sum x + b \sum x^2$

n implies the number of observations. From the above equation, we may get the value of y if the values of a, b, and x are known to us. We may also express the equation in another form which is shown below:

$y - \dot{Y} = b_{yx} (x - \dot{X})$

Here, b_{yx} is the regression coefficient. We can apply the following formula to determine the value of the regression coefficient.

$b_{yx} = [\sum xy/n - (\sum x/n) (\sum y/n)] / \sum x^2/n - (\sum x/n)^2$

There is also an alternative formula through which we may get the value of the coefficient. $b_{yx} = r(\partial y/\partial x)$

∂x implies the standard deviation of variable x and ∂y implies the standard deviation of variable y. r implies the value of the correlation coefficient.

2. **Regression Line of x on y:** The equation will be expressed in the form of x = a + by where a and b are two constants that are determined with the help of two normal equations.

$\sum x = na + b \sum y$

$\sum xy = a \sum y + b \sum y^2$

n implies a number of observations. From the above equation, we may get the value of y if the values of a, b, and x are known to us. We may also express the equation in another form which is shown below:

$x - \dot{X} = b_{xy} (y - \dot{Y})$

Here, b_{xy} is the regression coefficient. We can apply the following formula to determine the value of the regression coefficient.

$b_{xy} = [\sum xy/n - (\sum x/n)(\sum y/n)] / \sum y^2/n - (\sum y/n)^2$

There is also an alternative formula through which we may get the value of the coefficient. $b_{xy} = r(\partial x/\partial y)$

∂x implies the standard deviation of variable x and ∂y implies the standard deviation of variable y. r implies the value of the correlation coefficient.

Interpretation of Regression Coefficients: We know the general form of the equation when y depends on x is $y = a + bx$

Suppose the b value is 1.25. It implies that 1 unit change of the explanatory parameter results in 1.25 units of change in the dependent parameter.

7.7 Properties of Regression Coefficient

1. Symbolically $r = \sqrt{bxy * byx}$

   Proof

   $b_{yx} = (n\,\Sigma XY - \Sigma X \Sigma Y)/(n\,\Sigma X^2 - (\Sigma X)^2)$ 1

   Similarly,

   $b_{xy} = (n\,\Sigma XY - \Sigma X \Sigma Y)/(n\,\Sigma Y^2 - (\Sigma Y)^2)$ 2

   Also by the product moment formula of the correlation coefficient

   $r = (n\,\Sigma xy - \Sigma x\,\Sigma y)$

   $$r = \left(n\sum xy - \sum x \sum y\right) / \sqrt{\left(n\sum x^2 - \left(\sum x\right)^2\right)\left(n\sum y^2 - \left(\sum y\right)^2\right)}$$

   $\Rightarrow r^2 = (n\,\Sigma xy - \Sigma x\,\Sigma y)^2 / [(n\,\Sigma X^2 - (\Sigma X)^2)(n\,\Sigma Y^2 - (\Sigma Y)^2)]$ 3

   Multiplying equations 1 and 2 we get

   $b_{yx} \times b_{xy} = (n\,\Sigma XY - \Sigma X \Sigma Y)/(n\,\Sigma X^2 - (\Sigma X)^2) \times (n\,\Sigma XY - \Sigma X \Sigma Y) / (n\,\Sigma Y^2 - (\Sigma Y)^2)$

   $\Rightarrow b_{yx} \times b_{xy} = (n\,\Sigma xy - \Sigma x\,\Sigma y)^2 / [(n\,\Sigma X^2 - (\Sigma X)^2)(n\,\Sigma Y^2 - (\Sigma Y)^2)]$

   $\Rightarrow b_{yx} \times b_{xy} = r^2$

2. Out of two regression coefficients, if one value is greater than 1 then the other value must be less than 1 and vice-versa.

3. Each regression coefficient value must hold the same sign.

4. In the case of a negative regression coefficient, ‘r’ will likewise be negative, and in the case of a positive regression coefficient, ‘r’ will also be positive. For example, if bxy = -0.2 and byx = -0.8 then $r = -\sqrt{0.2 * 0.8} = -0.4$

5. Regression coefficient means are higher than correlation coefficient values.

6. Coefficient value depends only on scale change.

7.8 Simple and Multiple Regressions

Problem 1: Determine the x and y regression using the given information:

Σx = 24; **Σy** = 44; **Σxy** = 306; $\Sigma x^2 = 164$; $\Sigma y^2 = 574$; n = 4

Solution: Regression line of y on x

y = a + bx

$b = (n\,\Sigma xy - \Sigma x.\,\Sigma y)/\,(n\Sigma x^2 - (\Sigma x)^2)$

b = (4×306 - 24×44)/ (4×164 - 24×24)

b = 2.1

y-bar = 44/4 = 11; x-bar = 24/4 = 6

a = y-bar– b x-bar

a = 11-2.1x6

a = 1.6

Regression line of y on x

y= 1.6 + 2.1x

Regression line of x on y

Suppose it is given by

x= c + dy

Where $d= (n\,\Sigma xy - \Sigma x.\,\Sigma y)/\,(n\,\Sigma y^2 - (\Sigma y)^2)$

i.e., d = (4×306 - 24×44)/ (4×574 - 44×44)

=> d = 0.47

c = x-bar– d y-bar

c = 6-0.47x11 =0.83

Thus, the estimated regression line of x on y is

x = 0.83+0.47y

Problem 2: You are given the variance of x = 9. The regression equations are 8x – 10y + 66 = 0 and 40x – 18y = 214. Work out:

a) Mean of x and y
b) SD of y.

Solution: Regression equation of y on x

8x-10y +66=0

=> y = (8x+66)/10

=> y = 0.8x+6.6

=> Regression coefficient of y on x = $b_{yx} = 0.8$ ……1

Regression equation of x on y

40x -18y =214

=> x = (18y + 214)/40

=> x = 0.45y + 5.35

=> Regression coefficient of x on y = b_{xy} = 0.45 2

Correlation Coefficient r_{xy} = √0.8 × 0.45 = 0.6

$b_{yx} = r\sigma_y/\sigma_x$

0.8 = 0.6 x σ_y/3

σ_y = 4

Average values of x and y are the point of intersection of two lines of regressions

8x - 10y = - 66 => 4x - 5y = - 33 3

40x - 18y = 214 => 20x – 9y = 107 4

Solving equations 3 and 4

Average of x =13 and average of y = 17

Problem 3: The equations of two regression lines for the variables x and y are:

4x – 5y + 32 = 0 and

20x – 9y = 107

a) Categorize the regression lines
b) Find the r.
c) Find the mean of x and y.
d) Determine the estimated value of y, when x = 10.
e)

Solution: Two regression lines are given as

x on y

4x - 5y + 32 = 0

- 4x-5y =-32
- 4x = 5y-32
- x = 5/4y – 8 1

y on x

20x - 9y =107

- -9y = 107-20x
- 9y = 20x-107
- y= 20/9x-107/9 2

Since the product of the slope of y from equation 1 and slope of x from equation 2 = 5/4×20×9 = 100/36 which is greater than 1, equations 1 and 2 cannot be regression equations as r (b_{xy} × $\mathbf{b}_{yx}$) have to be less than 1

The regression line of y on x will be

4x - 5y + 32 = 0

- 5y = 4x+32
- y = 4/5x+32 3
- b_{yx} = 4/5

Regression line x on y

20x - 9y = 107

- 20x = 9y+107
- x= 9/20y+107/20 4
- b_{xy} = 9/20
- correlation coefficient =
- Correlation coefficient = 0.6

The two provided equations may be solved to get the averages of x and y.

4x – 5y + 32 = 0 and 20x – 9y = 107

20x-9y = 107

20x-25y = 160

16y = -53

y = - 53/16

x = 5/4(- 53/16) – 8

- **x = - 265/64 -8 = -12.14**

The estimated value of y when x = 10

y = 4/5x + 32

y = 4 × 10/5 + 32

y = 8 + 32 = 40

Problem 4: From the data given below, find:

i) The two regression equations
ii) The value of r
iii) Estimate Statistics marks when marks in Mathematics = 30

Marks in Mathematics	25	28	35	32	31	36	29	38	34	32
Marks in Statistics	43	46	49	41	36	32	31	30	33	39

Solution

Marks in mathematics (x)	Marks in statistics (y)	x2	y2	XV
25	43	625	1849	1075
28	46	"84	2116	1288
35	49	1225	2401	1715
32	41	1024	1681	1312
31	36	961	1296	1116
36	32	1296	1024	1152
29	31	841	961	899
38	30	1444	900	1140
34	33	1156	1089	1122
32	39	1024	1521	1248
320	380	10380	14838	12067

Mean of x = 320/10 = 32

Mean of y = 380/10 = 38

Regression line of x on y

$X = a + b_{xy}\ y$

b_{xy} = (n SXY - ΣXSY)/(n Sy^2 - (Σ y)^2)

- b_{xy} = (10X 12067 – 320X380)/(10X 14838-(380)^2)
- b_{xy} = (120670 – 121600)/148380-144400)
- b_{xy} = -930 / 3980
- **b_{xy} = -0.2336**

a = mean x – b_{xy} X mean y

a = 32 + 0.2336 X 38 = 33.8688

Therefore regression line of x on y

x = 33.8688 – 0.2336 y

Regression line of y on x

$y= a + b_{yx}\ x$

b_{yx} = (n Sxy - ΣxΣ y)/(n Sx^2 - (Σ x)^2)

- b_{yx} = (10X 12067 – 320X380)/(10X 10380-(320)^2)
- b_{yx} = (120670 – 121600)/103800-102400)
- b_{yx} = -930 / 1400
- **b_{yx} = -0.664**

a = mean y – b_{xy} X mean x

a = 38+0.664X 32 = 59.248

Therefore, the regression line of y on x is

y = 59.248 – 0.664 x

The correlation coefficient between the marks of mathematics and the marks of statistics is:

r_{xy} = -0.3938

Most likely marks in statistics when marks in mathematics = 30 are obtained from

Regression line of y on x

y = 59.248 – 0.664 x

y = 59.248 -0.664X 30 = 39.328

Estimated marks of statistics = 39.328

Problem 5: Calculate the Regression equation from the following information and also find the expected value of y where x=10.

Solution

X	1	2	3	4	5	$\sum x = 15$
Y	3	2	5	4	6	$\sum y = 20$
$X = x - \overline{x}$	-2	-1	0	1	2	
$Y = y - \overline{y}$	-1	-2	1	0	2	
X^2	4	1	0	1	4	$\sum x^2 = 10$
Y^2	1	4	1	0	4	$\sum y^2 = 10$
Xy	2	2	0	0	4	$\sum xy = 8$

$$\overline{x} = \frac{15}{5} = 3, \overline{y} = \frac{20}{5} = 4$$

$$byx = \frac{\sum xy}{\sum y^2} = \frac{8}{10} = .8, bx\,y = \frac{\sum xy}{\sum y^2} = \frac{8}{10} = .8$$

R.E y on x

$$y - \overline{y} = byx\,(x - \overline{x})$$

Y- 4 =.8(X-3) or, y-4 =.8x - 2.4 or, y = .8x + 1.6 ...... (1)

R.E of x on y

$X - \overline{x}\,bxy\,(y - y)\,or, x - 3 = .8(Y - 4)\,or, x - 3 = .8y - 3.2\,or, x = .8y - .2$ .... (2)

Here we have to find the value of y when x is given. So, the required equation is R.E of y on x.

Therefore the expected value of y

y = .8(10) + 1.6 = 9.6

Problem 6: Find the Regression equation from the following information.

x	6	09	10	12	14	$\sum x=51$
y	8	6	11	10	21	$\sum y=56$
	36	81	100	144	196	$\sum$
	64	36	121	100	441	$\sum=762$
xy	48	54	110	120	294	$\sum xy=626$

Solution

$$\bar{x}=\frac{51}{5}=10.2, \bar{y}=\frac{56}{5}=11.2, \text{cov}(xy)=\frac{\sum xy}{n}-\frac{\sum x}{n}.\frac{\sum y}{n}=\frac{626}{5}-10.2\times 11.2$$

$$=125.8-114.24=10.96$$

$$\rho x^2=\frac{\sum x^2}{n}-\left(\frac{\sum x}{n}\right)^2=\frac{557}{5}-(10.2)^2=111.4-104.04=7.36$$

$$\rho y^2=\frac{\sum y^2}{n}-\left(\frac{\sum y}{n}\right)^2=\frac{762}{5}-(11.2)^2=152.4-125.44=26.95$$

$$byx=\frac{\text{cov}(xy)}{\rho x^2}=\frac{10.96}{7.36}=1.5$$

$$byx=\frac{\text{cov}(xy)}{\rho x^2}=\frac{10.96}{26.96}=.41$$

Regression equation of y on x

$y-\bar{y}=byx(x-\bar{x})$ or, $y-11.2=1.5(x-10.2)$ or, $y-11.2=1.5x-15.3$ or, $y=1.5x-4.1$ ..(1)

Regression equation of x on y

$x-\bar{x}=bxy(y-\bar{y})$ 10.2 = .41(y-11.2) or, x-10.2 = .41y-4.592 or, x =.41y + 5.608

Problem 7: Calculate the Regression equation from the following information.

x	25	22	28	26	35	20	22	40	20	18	$\sum x=256$
y	18	15	20	17	22	14	16	21	15	14	$\sum y=172$
u=x-25	0	-3	3	1	10	-5	-3	15	-5	-7	$\sum u=6$
v=y-17	1	-2	3	0	5	-3	-1	4	-2	-3	$\sum v=2$
	0	9	9	1	100	25	9	225	25	49	$\sum u^2=452$
	1	4	9	0	25	9	1	16	4	9	$\sum v^2=79$
uv	0	6	9	10	50	15	3	60	10	21	$\sum uv=174$

Solution

$$x=\frac{256}{10}=25.6,\frac{172}{10}=17.2$$

$$\text{Cov}(uv)=\frac{\sum uv}{n}-\frac{\sum u}{n}.\frac{\sum v}{n}\frac{174}{10}-\frac{6}{3}.\frac{2}{10}=17.4-.12=17.28$$

$$\rho u^2=\frac{\sum u^2}{n}-\left(\frac{\sum u}{n}\right)^2=\frac{452}{10}-\left(\frac{6}{10}\right)^2 = 45.2-.36=44.84$$

$$\rho v^2=\frac{\sum v^2}{n}-\left(\frac{\sum v}{n}\right)2=\frac{78}{10}-\left(\frac{2}{10}\right)2=\frac{78}{10}-\left(\frac{2}{10}\right)^2 7.8.04=7.76$$

$$byx=\frac{\text{cov}(uv)}{\rho u^2}=\frac{17.28}{44.84}=.38$$

$$bxy=\frac{\text{cov}(uv)}{\rho v^2}=\frac{17.28}{7.78}=2.23$$

Regression equation of y on x

$y-\overline{y}=byx(x-\overline{x})$ or, y-17.2=.38(x-25.6) or, y-17.2 =.38x-9.728 or, y = .38x + 7.472

Regression equation of x on y

$x-x=bxy(y-\overline{y})$ or, x-25.6=2.23(y-17.2) or, x-25.6=2.23y-38.356or, x = 2.23y-12.756

Problem 8: In a distribution $\overline{x}=39.5\overline{y}=47.5\delta x=10.8\delta y16.8r=4$ Find out the regression equation.

Solution

$$byx=r.\frac{\delta y}{\delta x}=\frac{.4\times 16.8}{10.8}=.62$$

$$bxy=\frac{4\times 10.8}{16.8}=.26$$

Regression equation x on y

$x-\overline{x}=bxy(y-\overline{y})$ or, x-39.5=.26(y-47.5) or, x-39.5 =.26y-12.35 or, x =.26y + 27.15

Therefore R.E of y on x

$y-y=byx(x-\overline{x})$ or, y-47.5=.62(x-39.5) or, y-47.5=.62x-24.49 or, y =.62x + 2.301

Problem 9: In a distribution out of 10 pairs of observation mean is 8223 and the variance is 4816.

$\sum(x-\bar{x})(y-\bar{y})=64$

Calculate the R.E.

Solution

$$Cov(xy)=\frac{\sum(x-\bar{x})(y-\bar{y})}{n}=\frac{64}{10}=6.4$$

$n=10, \bar{x}=8, \bar{y}=23\ \delta x^2=4,\ \delta y^2=16$

$$byx=\frac{cov(xy)}{\delta x^2}=\frac{6.4}{4}=1.6\quad bxy=\frac{cov(xy)}{\delta y^2}=\frac{6.4}{16}=.4$$

R.E of y on x

$y-\bar{y}=byx(x-\bar{x})$ or, y-23 =1.6(x-8) or, y-23 =1.6-12.8 or, y =1.6x + 10.2

R.E of x on y

$x-\bar{x}=bxy(y-\bar{y})$ or, x-8=.4(y-23) or, x-8=.4y-9.2 or, x=.4y -1.2

Problem 10: Determine least square lines using – **i)** x as the independent and **ii)** y as the independent

Find the value of y when x =5 and also find x when y =5

x	3	5	6	8	9	11
y	2	3	4	6	5	8

Solution

x	y	x^2	y^2	xy
3	2	9	4	6
5	3	25	9	15
6	4	36	16	24
8	6	64	36	48
9	5	81	25	45
11	8	121	64	88
42	28	336	154	226

i) Where x is the independent variable:

Let the equation of the straight line be y = a + bx .....(1)

The normal equations are

$\sum y = na + b\sum x$ .....(2)

and $\sum xy = a\sum x + b\sum x^2$ .....(3)

From (2) $28 = 6a + 42b$ .....(4)

From (3) $226 = 42a + 336b$ .....(5)

From (4) and (5), we get a = -.33 and b = .71

From (1) the required equation is $y = -.33 + .71x$

When x =5, y = -.33 + .71 x 5 = 3.22

ii) Where y is the independent variable:

Let the equation of the straight line be $x = A + By$ .....(6)

The normal equations are

$\sum x = nA + B\sum y$ .....(7)

and $\sum xy = A\sum y + B\sum y^2$ .....(8)

From (7) $42 = 6A + 42B$ .....(9)

From (8) $226 = 28A + 154B$ .....(10)

From (9) and (10), we get A = 1 and B = 1.29

From (6) the required equation is $x = 1 + 1.29y$

When y = 7, x = 1 + 1.29 x 7 = 10.03

Problem 11: From the following data find the two regression equations:

x	1	2	3	4	5
y	2	3	5	4	6

Predict the value of y when x = 2.5.

Solution

Calculations for Regression Equations

x	y	x^2	y^2	xy
1	2	1	4	2
2	3	4	9	6
3	5	9	25	15
4	4	16	16	16
5	6	25	36	30
15	20	55	90	69

$\bar{x} = \sum x / n = 15/5 = 3$

$\bar{y} = \sum y / n = 20/5 = 4$

$b_{yx} = [\sum xy/n - (\sum x/n)(\sum y/n)] / \sum x^2/n - (\sum x/n)^2 = .9$

Similarly, $b_{xy} = [\sum xy/n - (\sum x/n)(\sum y/n)] / \sum y^2/n - (\sum y/n)^2 = .9$

The regression equation of y on x is $y - \bar{y} = byx(x - \bar{x})$

y = .9x + 1.3 ...... (1)

The regression equation of x on y is x- = b_{xy}(y -)

x = .9y - .6 .... (2)

When x = 2.5 from (1) we get y = .9 x 2.5 + 1.3 = 3.55

7.9 Concept of R Square in Regression

R-squared (R^2), also known as the coefficient of determination, is a statistical measure used in regression analysis to assess the goodness of fit of a regression model to the observed data. It provides information about the proportion of the variance in the dependent variable that is explained by the independent variables in the model. In other words, R-squared quantifies the percentage of variation in the dependent variable that is accounted for by the independent variables.

R-squared values range from 0 to 1, where:

0 indicates that the independent variables do not explain any of the variability in the dependent variable, and the model is a poor fit.

1 indicates that the independent variables perfectly explain all the variability in the dependent variable, and the model is an excellent fit.

In practice, R-squared values between 0 and 1 are more common, with higher values indicating better model fit. However, it is essential to remember that a high R-squared value does not necessarily mean that the model is a good predictor of future data or that it is a meaningful model. It is important to consider other factors and perform additional analyses to validate the model's suitability.

Here's an example to illustrate R-squared:

Suppose you are interested in predicting a person's salary (dependent variable) based on their years of experience (independent variable). You collect data on 100 individuals and fit a linear regression model to the data. After running the regression analysis, you obtain an R-squared value of 0.85. This R-squared value of 0.85 means that approximately 85% of the variability in salaries can be explained by the number of years of experience. In other words, the remaining 15% of the variability in salaries is attributed to other factors not included in the model, such as education, job role, location, etc.

7.10 Summary

Regression is applied for forecasting the value of an unknown variable with the help of a known variable.

A statistical method for determining the relations between dependent parameters and one or more independent parameters is called regression analysis. For forecasting purposes, regression is frequently utilized. Regression is used in many different disciplines, including management, business, engineering, and social science.

Regression analysis can be used in many different ways by an organization. Regression analysis aids an organization in better decision-making by enabling them to comprehend what their data points signify and make use of them accordingly.

Establishing a functional connection between two variables and utilizing that relationship to predict the future are the goals of regression analysis.

A statistical method for determining the relations between dependent parameters and 1 or more explanatory parameters is called regression analysis. For forecasting purposes, regression is frequently utilized. The dependent parameter is the one that is predicted, while the explanatory parameter is the one that serves as the foundation for the prediction. For instance, if we are interested in identifying the impact of price on demand then we may find the price for a certain demand and vice-versa.

7.11 Self-Assessment

1. What is regression analysis?
2. Discuss the assumptions of regression analysis.
3. Distinguish between simple linear regression and multiple regression.
4. What are the advantages of regression analysis?
5. How regression is different from correlation?
6. Why do we have two regression lines?
7. Establish the relation between regression coefficient and correlation coefficient.
8. State the features of the regression coefficient.
9. How can we compute the standard error of estimates in regression?
10. Discuss the drawbacks of correlation analysis.
11. Calculate the regression coefficient and obtain the lines of regression for the following data.

X	1	2	3	4	5	6	7
Y	9	8	10	12	11	13	14

12. Calculate the two regression equations of X on Y and Y on X from the data given below, taking deviations from actual means of X and Y.

Price (Rs.)	10	12	13	14	15	16
Demand	30	35	40	45	33	41

Estimate the likely demand when the price is Rs.20.

13. Find the means of X and Y variables and the coefficient of correlation between them from the following two regression equations:

$2Y–X–50 = 0$

$3Y–2X–10 = 0.$

14. In a laboratory experiment on correlation research study, the equation of the two regression lines was found to be $2X–Y+1=0$ and $3X–2Y+7=0$. Find the means of X and Y. Also work out the values of the regression coefficient and correlation between the two variables X and Y.

15. There are two series of index numbers P for price index and S for stock of the commodity. The mean and standard deviation of P are 100 and 8 and of S are 103 and 4 respectively. The correlation coefficient between the two series is 0.4. With these data obtain the regression lines of P on S and S on P.

Unit -VIII

Index Numbers

8.1 Learning Objectives

After completing this chapter, you should be able to:

i) Understand the concept and significance of index numbers;

ii) Identify different construction methods;

iii) Understand the importance of selecting appropriate base and current periods;

iv) Recognize the properties and tests of adequacy;

v) Recognize the challenges and limitations of index numbers;

vi) Stay informed of recent advances;

vii) Synthesize practical examples and calculations for creating index numbers.

8.2 Introduction

In this chapter, we embark on a journey through the intricate world of index numbers, a cornerstone in the realm of statistics. We will explore what index numbers are; their various types; and the methodologies employed in their construction. Delving into real-world applications, we'll uncover how these numerical tools are pivotal in measuring economic trends, informing policy decisions, and shaping our understanding of complex data sets. By the chapter's end, you'll gain insights into not only the mechanics but also the profound impact of index numbers in our data-driven society.

8.3 Definition and Overview of Index Number

Index numbers are statistical measures designed to show the changes in a variable or group of variables over time. They are typically expressed as a percentage of a base value, allowing for a comparison of relative changes over different periods. These numbers simplify the understanding of large data sets by condensing them into a single number. It is commonly used in economics and social sciences. Index numbers measure variations in prices, quantities, or values of commodities and services.

8.4 Importance of Index Number in Statistics

The significance of index numbers in statistics cannot be overstated. They are essential tools for interpreting complex economic data, providing a clear and concise way to track changes in economic indicators such as inflation, cost of living, industrial production, and many more. By converting vast amounts of data into a simple index, they allow policymakers, economists, and business leaders to make informed decisions based on trends and comparisons over time. In the realm of statistics, index numbers transform raw data into actionable insights, facilitating the analysis of economic and social phenomena.

8.5 Historical Background

The concept of index numbers dates back several centuries, with their use becoming more prevalent in the 19th and 20th centuries alongside the growth of modern economics and statistical science. One of the earliest known applications was in the late 1700s when economists began using index numbers to understand price variations and purchasing power. The development of the Consumer Price Index (CPI) and other similar indices in the early 20th century marked a significant advancement in this field. These indices have evolved over time, incorporating more sophisticated techniques and broader data to reflect economic realities more accurately. The historical evolution of index numbers mirrors the development of statistical methods and economic theory, highlighting their intertwined nature.

8.6 Base Period and Current Period

Understanding the concepts of the base year/period and current year/period is essential in the construction and interpretation of index numbers. These periods are fundamental in comparing changes over time in economic and statistical data.

8.6.1 Base Period Selection

The base period is a point of time used as a standard of comparison for other periods. The choice of a base year is crucial as it influences the interpretation of the index numbers. Key considerations include:

i) **Relevance:** The base period should be representative and typical of the conditions over the period being studied.

ii) **Stability:** Ideally, the base period should be free from any abnormal events, like economic crises or natural disasters that might skew the data.

iii) **Recency:** The base period should be recent enough to be relevant but not so recent that it lacks perspective on longer-term trends.

8.6.2 Base Period Adjustments for Changes over Time

Over time, changes in the market, economy, and consumer preferences can make the base period less representative. Adjustments might include:

i) **Updating the Basket:** In price indices, periodically updating the basket of goods and services to reflect current consumption patterns.

ii) **Seasonal Adjustments:** Accounting for regular seasonal variations to prevent them from distorting the index.

iii) **Quality Adjustments:** Adjusting for improvements or declines in product quality can affect price comparisons.

8.7 Methods of Constructing Index Numbers

The construction of index numbers is a critical process in statistics, involving various methodologies. Each method has its unique approach and application, depending on the data's nature and the intended use of the index. The two primary categories are Simple Index Numbers and Weighted Index Numbers, with the latter including several subtypes such as the Laspeyres Index, Paasche Index, and Fisher Index.

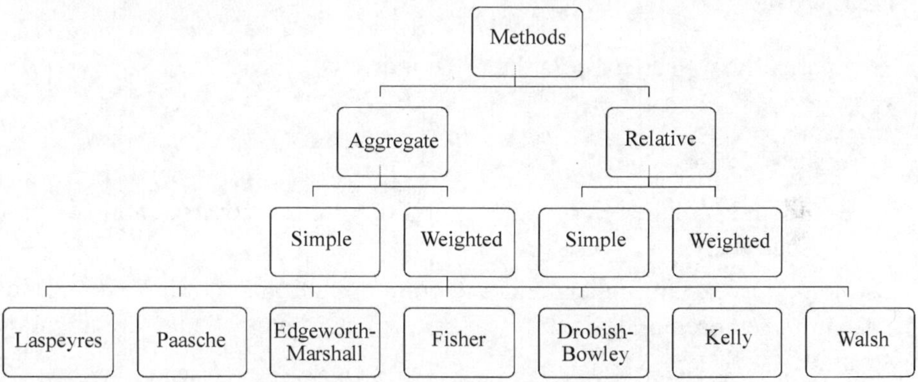

Symbols and notations to be used in the index functions:

p'_0, p''_0, p'''_0 ... represent prices in the base period denoted by the suffix 0.

q'_0, q''_0, q'''_0 ... represent quantities in the base period denoted by the suffix 0.

p'_n, p''_n, p'''_n ... represent prices in the given period denoted by suffix n.

q'_n, q''_n, q'''_n ... represent quantities in the given period denoted by the suffix n.

$p'_0, p''_0, p'''_0 + ...$ = Total of base period prices.

$q'_n, q''_n, q'''_n + ...$ = Total of base period prices.

Similarly, $\Sigma pn = p'_n + p''_n + p'''_n + ...$

$\Sigma pn = p'_n + p''_n + p'''_n + ...$

$\Sigma poqo = p'_0 q'_0 + p''_0 + p'''_0 q_0''' + ...$

$\Sigma pn\ qn = p'_n q'_n + p''_n q''_n + p_n''' + q_n''' + ...$

Here, p_{on} = Price index number for the given period n with respect to the base period 0.

q_{on} = Quantity index number for the given period n with respect to the base period 0.

I_{on} = Index number for the given period with respect to the base period 0.

8.7.1 Aggregate Method

8.7.1.1 Simple Index Numbers

Simple Index Numbers are the most straightforward method of constructing index numbers. They compare the value of a single item or a group of items in one period to the value in a base period. The formula is:

$$\text{Simple index number} = \frac{(\text{Value in current period})}{(\text{Value in base period})} \times 100$$

Mathematically, we can express the price index as:

$$I_{on} = \frac{\sum p_n}{\sum p_o} \times 100$$

Where and indicate the price of the commodity in the current period and base period respectively.

This method is easy to understand and compute but may not always provide an accurate representation if the quantities of the items change significantly over time.

Example: Suppose you want to calculate the simple price index of a product over a span of four years. The base year (the year against which you are comparing other years) is 2019. The prices of the product for each year are as follows:

Year	Price
2019	$50
2020	$55
2021	$60
2022	$65

Calculate the simple price index for each year from 2020 to 2022 with 2019 as the base year.

Solution

The simple price index is calculated using the formula:

$$I_{on} = \frac{\sum p_n}{\sum p_o} \times 100$$

Price Index Number for the year 2020 (I_{01}),

$$I_{01} = \frac{\sum p_1}{\sum p_o} \times 100 = \frac{55}{50} \times 100 = 100$$

For 2021,

$$I_{02} = \frac{\sum p_2}{\sum p_o} \times 100 = \frac{60}{50} \times 100 = 120$$

For 2022,

$$I_{03} = \frac{\sum p_3}{\sum p_o} \times 100 = \frac{65}{50} \times 100 = 130$$

Interpretation: The price index of 110 indicates that the price of the product in 2020 was 110% of its price in the base year (2019). This means that there was a 10% increase in price compared to 2019. The price index of 120 suggests a 20% increase in the price of the product from the base year. The index of 130 indicates a 30% increase in the price from 2019. The outcome shows a steady increase in the price of the product year over year, with each successive year experiencing a higher index number, reflecting rising prices compared to the base year of 2019.

8.7.1.2 Weighted Index Numbers

Weighted Index Numbers are more complex but offer a more accurate picture as they take into account the relative importance of different items. Mathematically, we can express a weighted aggregative index as:

$$I_{0n} = \frac{\sum p_n w}{\sum p_o w} \times 100$$

Where p_n and indicate the price of the commodity in the current period and base period. $w'\ w'', w'''$, ... indicate the weights of different items. So, $\Sigma pnw = p'_n w' + p_n'' w'' + p_n''' + w''' + ...$ and $\Sigma p_0 w = p'_0 w' + p_0'' w'' + ...$ Generally, the weights used in this method are quantities (q) consumed, marketed or sold in the base period or given period or the average of several periods.

1. Laspeyre's Index

The Laspeyre's Index measures price changes while keeping the quantity constant at the base period level. It is calculated as:

$$\text{Laspeyre's index} = \frac{\sum(\text{Current period prices} \times \text{base period quantities})}{\sum(\text{Base period prices} \times \text{base period quantities})} \times 100$$

Mathematically, substituting the equation can be expressed as:

$$I_{0n} \frac{\sum p_n q_0}{\sum p_0 q_0} \times 100$$

Example: Suppose you have a basket of two products: Apples and Bananas. The base year is 2019, and you want to calculate the Laspeyre's Price Index for 2021.

Year	Apple		Bananas	
	Price	Quantity	Price	Quantity
2019	$2	10	$1	15
2021	$3	10	$1.5	15

Solution

Items						
Apple	2	3	10	10	20	30
Bananas	1	1.5	15	15	15	22.5
Total					35	52.5

Laspeyre's Price Index

$$I_{0n} = \frac{\sum pnq_0}{\sum p_0 q_0} \times 100$$

$$I_{0n} = \frac{52.5}{35} \times 100 = 150$$

Interpretation: The Laspeyre's Price Index of 150 indicates that the cost of the same quantity of goods in 2021 is 150% of the cost in the base year (2019). This represents a 50% increase in the cost.

2. Paasche's Index

In contrast, the Paasche's Index uses the current period quantities to weight the prices. It is calculated as:

$$\text{Paasche's index} = \frac{\sum(\text{Current period prices} \times \text{current period quantities})}{\sum(\text{Base period prices} \times \text{current period quantities})} \times 100$$

Mathematically, it can be expressed as:

$$I_{0n} = \frac{\sum p_n q_n}{\sum p_0 q_n} \times 100$$

This index can capture consumers' substitution towards cheaper goods but may be more difficult to compute due to the need for current period quantity data.

Example: Let's continue with the example of Apples and Bananas. The base year is still 2019, and you want to calculate the Paasche's Price Index for 2021.

Year	Apple		Bananas	
	Price	Quantity	Price	Quantity
2019	$2	8	$1	20
2021	$3	8	$1.5	20

Solution

Items	p_0	p_n	q_0	q_n	$p_o q_n$	$p_n q_n$
Apple	2	3	8	8	16	24
Bananas	1	1.5	20	20	20	30
Total					36	54

Paasche's Price Index

$$I_{0n} = \frac{\sum p_n q_n}{\sum p_0 q_n} \times 100$$

$$I_{0n} = \frac{54}{36} \times 100 = 150$$

Interpretation: The Paasche's Price Index of 150 indicates that the cost of the current quantity of goods in 2021 is 150% of the cost of the same quantity in the base year (2019). This suggests a 50% increase in the cost, indicating inflation. The Paasche's Index reflects the current consumption behaviour more accurately since it considers the actual quantities bought in the current period, which may have adjusted due to price changes or other factors.

3. Drobish-Bowley Method

The Drobish-Bowley Method, also known as the Bowley's Index, is another approach to measure price changes over time. It's a compromise between the Laspeyre's and Paasche's indices, aiming to balance the perspectives of both. This method takes the average of the Laspeyre's and Paasche's indices to provide a middle-ground estimate of price changes.

The formula is:

$$\text{Drobish} - \text{bowley index} = \frac{\sum(\text{Current prices} \times \text{average quantities})}{\sum(\text{Base prices} \times \text{average quantities})} \times 100$$

Or, $\text{Drobish} - \text{bowley index} = \left[\frac{(\text{Laspeyres index} + \text{paasche index})}{2}\right]$

Mathematically, it can be expressed as:

$$I_{on} = \frac{\frac{\sum p_n q_0}{\sum p_0 q_0} + \frac{\sum p_n q_n}{\sum p_0 q_n}}{2}$$

This index offers a balanced view of price changes by accounting for both base period and current period quantities.

Example: Suppose we're analysing the price changes of two products, Coffee and Tea, from 2019 to 2021.

Year	Coffee		Tea	
	Price	**Quantity**	**Price**	**Quantity**
2019	\$4	10	\$2	20
2021	\$5	12	\$3	18

Solution

Items	p_0	p_n	q_0	q_n	p_0q_0	p_nq_0	p_0q_n	p_nq_n
Coffee	4	5	10	12	40	50	48	60
Tea	2	3	20	18	40	60	36	54
Total					80	110	84	114

Calculation of the Drobish-Bowley Index for 2021

$$I_{0n} = \frac{\frac{\sum p_n q_0}{\sum p_0 q_0} + \frac{\sum p_n q_n}{\sum p_0 q_n}}{2} \times 100$$

$$I_{0n} = \frac{\frac{110}{80} + \frac{114}{84}}{2} \times 100 = \frac{1.35 + 1.357}{2} \times 100 = 136.6$$

Alternatively, **Laspeyre's Index** = (110/80)*100 =137.5

Paasche's Index = (114/84)*100 = 135.7

Drobish-Bowley Index = (137.5+135.7)/2 = 136.6

Interpretation: The Laspeyre's Index of 137.5 suggests a 37.5% increase in the cost of the 2019 basket of goods when priced at 2021 prices. The Paasche's Index of 135.7 indicates a 35.7% increase in the cost of the 2021 basket of goods when priced at 2019 prices. The Drobish-Bowley Index of 136.6, being the average of the two, provides a balanced view, suggesting an overall price increase of approximately 36.6% from 2019 to 2021. This index offers a compromise between the two perspectives, considering both the changes in prices and quantities over the period.

4. Fisher's Index

The Fisher's Index is an attempt to combine the strengths of the Laspeyre's and Paasche's indices. It is the geometric mean of the two:

$$\text{Fisher's index} = \sqrt{\text{Laspeyres index} \times \text{paasche index}} \times 100$$

Mathematically, Fisher's Ideal formula can be expressed as:

$$\text{I0n} = \sqrt{\frac{\sum p_n q_0}{\sum p_0 q_0} \times \frac{\sum p_n q_n}{\sum p_0 q_n}} \times 100$$

This method is often considered the most accurate as it balances the strengths and weaknesses of the Laspeyre's and Paasche's methods.

Example: Construct the index number for 2021 using 2019 as the base year from the following data using the Fisher's Index.

Year	Product A		Product B	
	Price	Quantity	Price	Quantity
2019	$10	100	$20	50
2021	$12	110	$22	55

Solution

Items	p_0	p_n	q_0	q_n	p_0q_0	p_nq_0	p_0q_n	p_nq_n
Product A	10	12	100	110	1000	1200	1100	1320
Product B	20	22	50	55	1000	1100	1100	1210
Total					2000	2300	2200	2530

$$I_{0n} = \sqrt{\frac{\sum p_n q_0}{\sum p_0 q_0} \times \frac{\sum p_n q_n}{\sum p_0 q_n}} \times 100$$

$$I_{0n} = \sqrt{\frac{2300}{200} \times \frac{2530}{2200}} \times 100 = \sqrt{1.15 \times 1.15} \times 100 = 115$$

Interpretation: The Fisher's Index value of 115 indicates a 15% increase in the overall price level of these products from Year 1 to Year 2. This index effectively balances the two perspectives – one focusing on the base year's quantities (Laspeyre's) and the other on the current year's quantities (Paasche's) – providing a more comprehensive view of price changes over time.

5. Marshall-Edgeworth Method

The Marshall-Edgeworth Index modifies the Drobish-Bowley method by using a different average for quantities. It uses the arithmetic mean of the sums of the quantities instead of the quantities themselves. The formula is:

$$I_{0n} = \frac{\sum p_n (q_0 + q_n)/2}{\sum p_0 (q_0 + q_n)/2} \times 100 = \frac{\sum p_n q_0 + \sum p_n q_n}{\sum p_0 q_0 + \sum p_0 q_n} \times 100$$

Example: Compute Marshall-Edgeworth's index based on the following data considering 2022 as a base year:

Items	Quantity		Price	
	2022	2023	2022	2023
A	9	11	12	15
B	5	10	8	10
C	12	16	10	12

Solution

Items	p_0	p_n	q_0	q_n	p_0q_0	p_nq_0	p_0q_n	p_nq_n
A	12	15	9	11	108	135	132	165
B	8	10	5	10	40	50	80	100
C	10	12	12	16	120	144	160	192
Total					268	329	372	457

Calculation of Index number

$$I_{0n} = \frac{\sum p_n q_0 + \sum p_n q_n}{\sum p_0 q_0 + \sum p_0 q_n} \times 100$$

$$I_{0n} = \frac{329 + 457}{268 + 372} \times 100 = \frac{786}{640} \times 100 = 122.81$$

Interpretation: The Marshall-Edgeworth's Index value of 122.81 indicates a 22.81% increase in the overall price level of these products from 2022 to 2023.

6. Kelly's Method

Kelly's method is similar to the Marshall-Edgeworth method but uses geometric means instead of arithmetic means for the quantities. The formula for the price index is:

$$I_{0n} = \frac{\sum p_n q}{\sum p_0 q} \times 100$$

Where $q = \frac{q_0 + q_n}{2}$

Example: Construct Kelly's index based on the following data:

Items	Price		Quantity	
	Base year	**Current year**	**Base year**	**Current year**
A	20	25	10	12
B	30	35	15	18

Solution

Items	p_0	p_n	q_0	q_n	q	$p_n q$	$p_0 q$
A	20	25	10	12	11	275	220
B	30	35	15	18	16.5	577.5	495
Total						852.5	715

Kelly's Index calculation

$$I_{0n} = \frac{\sum p_n q}{\sum p_0 q} \times 100 = \frac{852.5}{715} \times 100 = 119.23$$

Interpretation: This result indicates that, on average, considering the average quantities of the base and current periods, the price level has increased by about 19.23% from the base period to the current period. This reflects the change in the overall price movement for the basket of goods when taking into account the average quantities of goods over the two periods.

7. Walsh's Method

Walsh's index is another variation that uses the geometric mean of prices and the geometric mean of quantities. The Walsh's price index formula is:

$$P_w = \frac{\sum p_n \sqrt{q_0 q_n}}{\sum p_0 \sqrt{q_0 q_n}} \times 100$$

Example: Construct Walsh's Price Index based on the following data:

Items	Price		Quantity	
	Base year	**Current year**	**Base year**	**Current year**
A	20	25	10	12
B	30	35	15	18

Solution

Items	p0	pn	q0	qn	$\sqrt{q_0 q_n}$	$p_n \sqrt{q_0 q_n}$	$p_0 \sqrt{q_0 q_n}$
A	20	25	10	12	10.95	273.75	219
B	30	35	15	18	16.43	575.05	492.9
Total						848.8	711.9

Calculation of Index number based on Walsh's formula:

$$P_w = \frac{\sum p_n \sqrt{q_0 q_n}}{\sum p_0 \sqrt{q_0 q_n}} \times 100 = \frac{848.8}{711.9} \times 100 = 119.23$$

Interpretation: The result indicates a 19.23% increase in price level from the base period to the current period when considering the geometric mean of the quantities produced in both periods. This measurement provides a balanced view of price changes, accounting for variations in both quantities and prices over time.

8.7.2 Relative Method

8.7.2.1 Simple Average of Price Relative Method

It is the mean of a group of price relatives that are derived by defining the prices of the specified period as a percentage of the base prices.

- When Arithmetic Mean (A.M.) is used:

$$I_{0n} = \frac{\sum \frac{p_n}{p_0}}{N} \times 100$$

- When Geometric Mean (G.M.) is used:

$$I_{0n} = \left(\frac{p'_n}{p'_0} \times \frac{p''_n}{p''_0} \times \frac{p'''_n}{p'''_0} \ldots \right)^{\frac{1}{N}} \times 100$$

Where N is the number of items/commodities.

Example: Construct a Price Index Number using a simple average of relative price method based on A.M. and G.M. techniques from the following data.

Items	Price	
	Base year	**Current year**
A	26	31
B	21	23
C	31	34
D	13	16
E	92	99

Solution

Items	p_0	p_n	$\frac{p_n}{p_0}$	log p_0	log p_n	log p_n - log p_0
A	26	31	1.192	1.415	1.491	0.076
B	21	23	1.095	1.322	1.362	0.040
C	31	34	1.097	1.491	1.531	0.040
D	13	16	1.231	1.114	1.204	0.090
E	92	99	1.076	1.964	1.996	0.032
Total			5.691			0.278

Using A.M., Simple average price relative index number is:

$$I_{0n} = \frac{\sum \frac{p_n}{p_0}}{N} \times 100 = \frac{5.691}{5} \times 100 = 113.82$$

Using G.M., the Simple average price relative index number is:

$$I_{0n} = \left(\frac{p_n^{'}}{p_0^{'}} \times \frac{p_n^{''}}{p_0^{''}} \times \frac{p_n^{'''}}{p_0^{'''}} \times \ldots \right)^{\frac{1}{N}} \times 100$$

Or, log $I = \frac{1}{5} \sum \log \frac{p_n}{p_0} + \log 100$

$$= \frac{1}{5} \sum (\log p_n - \log p_0) + \log 100$$

$$= \frac{1}{5} \times 0.278 + 2 = 0.0556 + 2 = 2.0556$$

$I_{0n} = 113.658[10^{\wedge}2.0556]$

8.7.2.2 Weighted Average of Price Relative Method

The weighted average of price relatives is calculated by multiplying each relative by its corresponding weight and then dividing the result by the sum of the weights (*w'*, *w''*, *w'''*, ...). Mathematically it can be expressed as:

$$I_{0n} = \frac{\sum \frac{p_n}{p_0} \times w}{\sum w} \times 100$$

Typically, the weights in this case are value weights. When the weights are base period values, then and we obtain from the above equation,

$$I_{0n} = \frac{\sum \frac{p_n}{p_0} \times p_0 q_0}{\sum p_0 q_0} \times 100 = \frac{\sum p_n q_0}{\sum p_0 q_0} \times 100$$

It is just the Laspeyre's formula.

Similarly, when weights are values of given period quantities at base prices, then and from the weighted average price relative method equation we obtain,

$$I_{0n} = \frac{\sum \frac{p_n}{p_0} \times p_0 q n}{\sum p_0 q n} \times 100 = \frac{\sum p_n q_n}{\sum p_0 q_n} \times 100$$

Which is Paasche's formula.

Example: Based on the below table, calculate the price index number for the year 2023 while considering 2021 as the base year applying a weighted average of price relative method.

Items	**Price**		**Weights**
	2021	**2023**	
Rice	170	190	11
Wheat	130	160	5
Pulse	470	490	4
Fish	15	18	3

Solution

Items	p_0	p_n	w	$\frac{p_n}{p_0}$	$\frac{p_n}{p_0} w$
Rice	170	190	11	1.118	12.298
Wheat	130	160	5	1.231	6.155

Pulse	470	490	4	1.042	4.168
Fish	15	18	3	1.2	3.6
Total			23	4.591	26.221

Weighted Average of Price Relative Index Number:

$$I_{0n} = \frac{\sum \frac{p_n}{p_0} \times w}{\sum w} \times 100$$

$$I_{0n} = \frac{26.221}{23} \times 100 = 114.004$$

8.8 Chain Base Index Numbers

Chain base index numbers address the issue of an outdated base period. Unlike fixed-base index numbers, which always compare to a single base period, chain base indices link successive periods together. Key aspects are:

i) **Flexibility:** They allow the base period to be updated regularly, typically annually.

ii) **Comparability:** By chaining, they provide a more accurate reflection of long-term changes.

Calculation: Each period is compared to the previous one, and these relative changes are linked (or 'chained') together to form a continuous series. Symbolically,

$I'_{0n} = I_{01} \times I_{12} \times \ldots \times I_{(n-1)}$; (omitting the factor 100)

Where I_{ij} represents the Index number with i as the base and j as the given year. $I_{01}, I_{12}, I_{23}, \ldots$ are called link indices and is called the Chain Index.

Omitting the factor 100, If we apply the Laspeyres formula then we obtain,

$$I_{01} = \frac{\sum p_1 q_0}{\sum p_0 q_0}; I_{12} = \frac{\sum p_2 q_1}{\sum p_1 q_1}; \ldots; I_{(n-1)}n = \frac{\sum p_n q_n - 1}{\sum p_n - 1 q_n - 1}$$

and,

$$I'_{0n} = \frac{\sum p_1 q_0}{\sum p_0 q_0} \times \frac{\sum p_2 q_1}{\sum p_1 q_1} \times \ldots \times \frac{\sum p_n q_{n-1}}{\sum p_{n-1} q_{n-1}}$$

Similarly, omitting the factor 100, if we apply Paasche's formula then we obtain,

and,

$$I_{01} = \frac{\sum p_1 q_1}{\sum p_0 q_1}; I_{12} = \frac{\sum p_2 q_2}{\sum p_1 q_2}; \ldots; I_{(n-1)n} = \frac{\sum p_n q_n}{\sum p_{n-1} q_n}$$

However, if we consider factor 100 then we have

$$I'_{01} = I_{01}$$

$$I'_{02} = \frac{I_{01} \times I_{12}}{100} = \frac{I'_{01} \times I_{12}}{100}$$

$$I'_{03} = \frac{I_{01} \times I_{12} \times I_{23}}{100 \times 100} = \frac{I'_{02} \times I_{34}}{100}$$

$$I'_{04} = \frac{I_{01} \times I_{12} \times I_{23} \times I_{34}}{100 \times 100 \times 100} = \frac{I'_{03} \times I_{34}}{100}$$

Chain base index numbers are particularly useful in fast-changing environments where a fixed base period might quickly become irrelevant. However, they can be more complex to calculate and interpret than fixed-base index numbers.

Example: Compute chain index numbers for the years 2019 to 2023 considering 2018 as the base year (Base 2018 100).

YEAR	2019	2020	2021	2022	2023
LINK INDEX	102	97	106	113	110

Solution

Calculation of Chain Index Numbers (Base 2018 100)

Year	Link Index	Chain Index
2018		100
2019		$I'_{01} = 102$
2020		$I'_{02} = \frac{102 \times 97}{100} = 98.94$
2021		$I'_{03} = \frac{98.94 \times 106}{100} = 104.88$
2022		$I'_{04} = \frac{104.88 \times 113}{100} = 118.51$
2023		$I'_{05} = \frac{118.51 \times 110}{100} = 130.36$

8.9 Quantity Index Number

A quantity index number measures the change in volume of quantity produced, consumed or distributed. Problems and methods of construction of this index number are similar to those involved in the price index. So, this will not be discussed separately. The quantity index number formula can be obtained easily by changing p to q in the corresponding price index number formula. A glimpse of the price and quantity index number formula is presented below:

Method	Price Index Number ()	Quantity Index Number ()
	1. Aggregate Method	
a) Simple	$\frac{\sum p_n}{\sum p_0}\times 100$	$\frac{\sum q_n}{\sum q_0}\times 100$
b) Weighted	$\frac{\sum p_n w}{\sum p_0 w}\times 100$	$\frac{\sum q_n w}{\sum q_0 w}\times 100$
i) Laspeyres	$\frac{\sum p_n q_0}{\sum p_0 q_0}\times 100$	$\frac{\sum q_n p_0}{\sum q_0 p_0}\times 100$
ii) Paasche	$\frac{\sum p_n q_n}{\sum p_0 q_n}\times 100$	$\frac{\sum q_n p_n}{\sum q_0 p_n}\times 100$
iii) Marshall- Edgeworth	$\frac{\sum p_n q_0 + \sum p_n q_n}{\sum p_0 q_n + \sum p_0 q_n}\times 100$	$\frac{\sum q_n p_0 + \sum q_n p_n}{\sum q_0 p_0 + \sum q_0 p_n}\times 100$
iv) Fisher	$\sqrt{\frac{\sum p_n q_0}{\sum p_0 q_0}\times\frac{\sum p_n q_n}{\sum p_0 q_n}}\times 100$	$\sqrt{\frac{\sum q_n p_0}{\sum q_0 p_n}\times\frac{\sum q_n p_n}{\sum q_0 p_n}}\times 100$
	2. Relative Method	
a) Simple	$\frac{\sum \frac{p_n}{p_0}}{N}\times 100$	$\frac{\sum \frac{q_n}{q_0}}{N}\times 100$
b) Weighted	$\frac{\sum \frac{p_n}{p_0}\times w}{\sum w}\times 100$	$\frac{\sum \frac{q_n}{q_0}\times 100}{\sum w}\times 100$

8.10 Properties of Good Index Numbers

A well-constructed index number is a powerful tool in economic and statistical analysis. For an index number to be effective, it should possess certain key properties and pass specific tests of adequacy. It's also important to understand its limitations and potential biases.

1. Tests of Adequacy

a) **Time Reversal Test:** This test checks whether reversing the time order (swapping the base and current periods) leads to a reciprocal value of the original index. It ensures that the index is symmetrical over time. Symbolically,

$$I_{0n} = \frac{1}{I_{n0}} \text{ or, } I_{0n} \times I_{n0} = 1 \text{ (omitting the factor 100)}$$

This test is satisfied by simple aggregate index, simple G.M. of price relatives formula, weighted G.M. of price relative formula with constant weights, Marshall Edgeworth's formula and Fisher's formula.

Example 1: Prove that unweighted G.M. of price relatives' formula satisfies the time reversal test.

Solution

For unweighted G.M. of price relatives (omitting factor 100) we have,

$$I_{0n} = \left(\frac{p_n'}{p_0'} \times \frac{p_n''}{p_0''} \times \frac{p_n'''}{p_0'''} \times \cdots\right)^{\frac{1}{N}} \times 100 \text{ and } I_{n0} = \left(\frac{p_0'}{p_n'} \times \frac{p_0''}{p_n''} \times \frac{p_0'''}{p_n'''}\right)^{\frac{1}{N}} \times 100$$

So, $I_{0n} \times I_{no} = 1$

Thus, the unweighted G.M. of price relative's formula satisfies the time reversal test.

Example 2: Prove that the time reversal test is neither obeyed by Paasche's method nor Laspeyres' method.

Solution

Omitting factor 100, from Paasche's method we obtain:

$$I_{0n} = \frac{\sum p_n q_n}{\sum p_0 q_n} \text{ and } I_{n0} = \frac{\sum p_0 q_0}{\sum p_n q_0}$$

But, $I_{0n} \times I_{n0} = \frac{\sum p_n q_n}{\sum p_0 q_n} \times \frac{\sum p_0 q_n}{\sum p_n q_0}$ is not in general equal to 1.

Thus, Paasche's method does not satisfy the time reversal test.

Similarly, by omitting factor 100, from Laspeyres' method we obtain:

$$I_{0n} = \sqrt{\frac{\sum p_n q_0}{\sum p_0 q_0} \times \frac{\sum p_n q_n}{\sum p_0 q_n}} \text{ and } I_{n0} = \sqrt{\frac{\sum p_0 q_n}{\sum p_n q_n} \times \frac{\sum p_0 q_0}{\sum p_n q_0}}$$

But, $I_{0n} \times I_{n0} = \frac{\sum p_n q_0}{\sum p_0 q_0} \times \frac{\sum p_0 q_n}{\sum p_n q_n}$ is not in general equal to 1.

Thus, Laspeyres' method also does not satisfy the time reversal test.

Example 3: Prove that Fisher's Index Number formula satisfies the time reversal test.

Solution

Omitting the factor 100 and applying the Fisher's index Number formula, we get

$$I_{0n} = \sqrt{\frac{\sum p_n q_0}{\sum p_0 q_0} \times \frac{\sum p_n q_n}{\sum p_0 q_n}} \text{ and } I_{n0} = \sqrt{\frac{\sum p_0 q_n}{\sum p_n q_n} \times \frac{\sum p_0 q_0}{\sum p_n q_0}}$$

So, $I_{0n} \times I_{n0} = 1$

Hence, Fisher's Index Number formula satisfies the time reversal test.

b) **Factor Reversal Test:** This test assesses if multiplying the price index by a quantity index for the same items and periods gives the value change in total. It ensures that the index remains valid when factors like price and quantity are interchanged. Symbolically,

$$P_{0n} \times Q_{0n} = V_{0n}; \text{where } V_{0n} = \frac{\sum p_n q_n}{\sum p_0 q_0}$$

This test is satisfied only by Fisher's formula.

Example: Prove that Fisher's Index Number formula satisfies the factor reversal test.

Solution

According to Fisher's Index Number formula omitting the factor 100, we have,

$$P_{0n} = \sqrt{\frac{\sum p_n q_0}{\sum p_0 q_0} \times \frac{\sum p_n q_n}{\sum p_0 q_n}} \text{ and } Q_{0n} = \sqrt{\frac{\sum q_n p_0}{\sum q_0 p_0} \times \frac{\sum q_n p_n}{\sum q_0 p_n}}$$

$$P_{0n} \times Q_{0n} = \sqrt{\frac{\sum p_n q_0}{\sum p_0 q_0} \times \frac{\sum p_n q_n}{\sum p_0 q_n} \times \frac{\sum q_n p_0}{\sum q_0 p_0} \times \frac{\sum q_n p_n}{\sum q_0 p_n}} = \frac{\sum p_n q_n}{\sum p_0 q_0} = V_{0n}$$

So, Fisher's Index Number formula satisfies the factor reversal test.

c) **Circular Test:** This is an extension of the time reversal test. An index number formula is said to satisfy this test if

$I_{01} \times I_{12} \times I_{23} \ldots \times I_{(n-1)n} \times I_{n0} = 1$; (*omitting factor* 100)

This test is satisfied only by the G.M. of relatives and by fixed-weight aggregates.

Example: Prove that the Fisher formula does not satisfy the circular test.

For Fisher we have

$$I_{01} = \sqrt{\frac{\sum p_1q_0}{\sum p_0q_0} \times \frac{\sum p_1q_1}{\sum p_0q_1}}, I_{12} = \sqrt{\frac{\sum p_2q_1}{\sum p_1q_1} \times \frac{\sum p_2q_2}{\sum p_1q_2}} \text{ and } I_{20} = \sqrt{\frac{\sum p_0q_2}{\sum p_2q_2} \times \frac{\sum p_0q_0}{\sum p_2q_0}}$$

$$I_{01} \times I_{12} \times I_{20} = \sqrt{\frac{\sum p_1q_0}{\sum p_0q_0} \times \frac{\sum p_1q_1}{\sum p_0q_1} \times \frac{\sum p_2q_1}{\sum p_1q_1} \frac{\sum p_2q_2}{\sum p_1q_2} \times \frac{\sum p_0q_2}{\sum p_2q_2} \times \frac{\sum p_0q_0}{\sum p_2q_0}}$$

$$= \sqrt{\frac{\sum p_1 q_0 \times \sum p_2q_1 \times \sum p_0q_2}{\sum p_0 q_1 \times \sum p_1q_2 \times \sum p_2q_0}} \neq 1$$

Hence, the Fisher formula does not satisfy the circular test.

Example: Using the following data shows that the Laspeyre's Price Index formula does not satisfy the time and factor reversal tests while Fisher's ideal formula satisfies both time and factor reversal tests.

Items	Base year		Current year	
	Price	Quantity	Price	Quantity
A	7	50	10	56
B	3	100	2	120
C	4	60	6	60
D	10	30	12	24
E	8	40	12	36

Solution

Items	p_0	p_n	q_0	q_n	p_0q_0	pnq_0	p_0q_0	p_nq_n
A	7	10	50	56	350	500	392	560
B	3	2	100	120	300	200	360	240
C	4	6	60	60	240	360	240	360
D	10	12	30	24	300	360	240	288
E	8	12	40	36	320	480	288	432
Total					1510	1900	1520	1880

Omitting the factor 100 from each index,

Laspeyres Index Number

$$I_{0n}=\frac{\sum p_nq_0}{\sum p_0q_0}=\frac{1900}{1510}=1.258 \text{ and } I_{n0}=\frac{\sum p_0q_n}{\sum p_nq_n}=\frac{1520}{1880}=0.809$$

$I_{0n} \times I_{no} = 1.258 \times 0.809 \neq 1$

Hence, the Laspeyres Index does not satisfy the time reversal test.

Again, for Laspeyres: $P_{0n}=\frac{\sum p_nq_0}{\sum p_0q_0}=\frac{1900}{1510}$ and $Q_{0n}=\frac{\sum q_np_0}{\sum q_0p_0}=\frac{1520}{1510}$

$$P_{0n}\times Q_{0n}=\frac{1900}{1510}\times\frac{1520}{1510} \text{ and } V_{0n}=\frac{\sum p_nq_n}{\sum p_0q_0}=\frac{1880}{1510}$$

Thus, $P_{0n} \times Q_{0n} \neq V_{0n}$

Hence, the Laspeyres formula does not satisfy the factor reversal test.

Fisher's ideal price index number

$$I_{0n}=\sqrt{\frac{\sum p_0q_n}{\sum p_0q_0}\times\frac{\sum p_nq_n}{\sum p_0q_n}}=\sqrt{\frac{1900}{1510}\times\frac{1880}{1520}}$$

$$I_{0n}=\sqrt{\frac{\sum p_0q_n}{\sum p_nq_n}\times\frac{\sum p_0q_0}{\sum p_nq_0}}=\sqrt{\frac{1520}{1880}\times\frac{1510}{1900}}$$

$$I_{0n}\times I_{n0}=\sqrt{\frac{1900}{1510}\times\frac{1800}{1520}}=\sqrt{\frac{1520}{1880}\times\frac{1510}{1900}}$$

Thus, Fisher's ideal index number formula satisfies the time reversal test.

Again, Fisher's Ideal Price Index Number $=\sqrt{\frac{\sum p_nq_0}{\sum p_0q_0}\times\frac{\sum p_nq_n}{\sum p_0q_n}}=\sqrt{\frac{1900}{1510}\times\frac{1880}{1520}}$

And Fisher's Ideal Quantity Index Number

$$(Q_{0n})=\sqrt{\frac{\sum q_np_0}{\sum q_0p_0}\times\frac{\sum q_np_n}{\sum q_0p_n}}=\sqrt{\frac{1520}{1516}\times\frac{1880}{1900}}$$

$$P_{0n}\times Q_{0n}=\sqrt{\frac{1900}{1510}\times\frac{1880}{1520}\times\frac{1520}{1510}\times\frac{1880}{1900}}=\frac{1880}{1510}=\frac{\sum p_nq_n}{\sum p_0q_0}=V_{0n}$$

Thus, Fisher's Ideal Index formula satisfies the factor reversal test.

1. Reliability and Validity

Reliability: A reliable index number produces consistent results under consistent conditions. It's crucial to ensure that the index is dependable over time and across different contexts.

Validity: An index number is valid if it accurately measures what it is intended to measure. For instance, a Consumer Price Index should accurately reflect changes in the cost of living.

2. Limitations and Biases

Selection Bias: The choice of items and their weights in the index can introduce bias, especially if they do not represent the broader market or population accurately.

Base Year Bias: The choice of the base year can significantly impact the index, especially if the base year is not typical of longer-term conditions.

Quality Adjustment: Difficulty in adequately adjusting for quality changes in goods and services can lead to inaccuracies in measuring true price changes.

Substitution Bias: Fixed-weight indices can fail to account for changes in consumption patterns, leading to overestimation or underestimation of price changes.

Incorporating these properties, tests, and an awareness of potential limitations and biases, is essential for creating and interpreting index numbers effectively. They ensure that index numbers are not only statistically sound but also meaningful and relevant in real-world applications.

8.11 Different Types of Index Numbers

Index numbers come in various forms, each tailored to measure different aspects of economic and social data. Understanding the types of index numbers is crucial for interpreting their implications correctly. The three main categories are Price Index Numbers, Quantity Index Numbers, and Special Purpose Index Numbers.

8.11.1 Price Index Numbers

Price Index Numbers are perhaps the most well-known category of index numbers. They measure the average change in prices of a basket of goods and services over time.

A. Consumer Price Index (CPI)

CPI is a vital indicator of the inflation rate in an economy. It tracks the changes in the prices of a basket of consumer goods and services, such as food, transportation, and medical care. CPI is often used to adjust salaries, pensions, and government benefits to maintain purchasing power in the face of inflation. The formula of CPI is as follows:

$$CPI = \frac{\text{Value of basket in current year}}{\text{Value of basket in base year}} \times 100$$

Here, The CPI calculation utilizes a composite basket of goods and services that consists of prominent commodities that are often purchased, particularly familiar in America.

Example: Construct CPI for 2023 based on the below data using 2015 as the base year.

Year	Basket value
2015	$500
2023	$650

Solution

$$CPI = \frac{650}{500} \times 100 = 130$$

Interpretation: The CPI of 130 in 2023 indicates that the cost of the basket of goods and services has increased by 30% from 2015 to 2023.

B. Cost of Living Index (CLI)

The Cost of Living Index is primarily designed to track the fluctuation in prices that consumers face for a predetermined basket of goods and services over a specific period. This index serves as an indicator of the varying costs required to maintain a comparable standard of living for a distinct demographic, typically characterized by stable consumption habits, residing in a region where retail prices show minimal disparity.

Classification of Items in the Cost of Living Index

The Cost of living index typically encompasses a diverse array of products and services, which are broadly categorized into five primary segments: **(1)** Food, **(2)** Apparel/Clothing, **(3)** Fuel and Light, **(4)** Housing, and **(5)** Miscellaneous Expenses.

Each category encompasses a wide range of specific items. For instance, the 'Food' category includes staples and commodities such as Rice, Wheat, Milk, various Vegetables, Eggs, Fish, Meat, Tea, Butter, and Ghee, among others. The selection of these consumption items is based on the typical consumption patterns of the target demographic for the index. To streamline the process and manage resources efficiently, it's crucial to select a limited yet representative assortment of items. Regular collection of retail price data for these items is essential, and it should reflect the actual sources from where the consumers typically purchase. In cases where price data varies across different sources, an average of these prices should be utilized.

Weighting and Calculation

Recognizing the differing relevance of various goods and services across diverse consumer segments, it's essential to integrate appropriate weighting in the calculation of the Cost of Living Index. For each of the primary categories previously mentioned, the index is calculated through one of several recognized techniques. These techniques include:

i) Aggregate Expenditure Method or Aggregate Method

ii) Family Budget Method or Method of Weighted Relatives

i) Aggregate Expenditure Method

This approach involves using the quantities or proportions of items consumed by a specific group in the base year as the weights. The index number is determined by the sum of the weighted prices in the current year as a percentage of those in the base year. Mathematically, it is represented as:

$$\text{Group index (I)} = \frac{\sum p_n q_n}{\sum p_0 q_0} \times 100$$

This formula aligns with the Laspeyres' method, previously discussed.

ii) Family Budget Method

Here, the focus is on estimating the expenditure of an average family on various items, based on a comprehensive analysis of family budgets for a large sample size, specifically targeting the demographic for which the index is being compiled. This investigation, conducted in the base year, utilizes random sampling techniques. The average family's expenditure on each item then becomes the weight (*w*) of that item within its respective category. Thus, the weights are the value weights (p_0 and q_0), and the weighted average of price relatives is used to determine the index number. The formula for this method is as follows:

$$\text{Group index (I)} = \frac{\sum \frac{p_n}{p_0} \times w}{\sum w} \times 100 = \frac{\sum \frac{p_n}{p_0} \times p_0 q_0}{\sum w} \times 100 = \frac{\sum p_n q_0}{\sum p_0 q_0} \times 100 \left[\text{Since } w = p_0 q_0\right]$$

Here, w represents the value weights, which are the product of the base year prices and quantities (p_0 and q_0). Notably, this method essentially mirrors the Aggregate method.

In practical applications, the expenditure of an average family on each item is often expressed as a percentage of the total expenditure for that category. These percentages are then utilized as the weights for each item. Consequently, the Group Index (*I*) is calculated as:

$$w = \frac{p_0 q_0}{\sum p_0 q_0} \times 100$$

So, Group Index $= \dfrac{\sum \frac{p_n}{p_0} \times w}{100} \times 100 = \sum \dfrac{p_n}{p_0} \times w$

Finally, the overall Cost of Living Index number (C.L.I.) is computed using the Weighted Arithmetic Average of the Group Indices:

$$\text{C.L.L.} = \frac{\sum IW}{\sum W}$$

In this formula, *W*, the weight of each group index, is determined by the percentage of the total expenditure an average family allocates to that particular category.

Example: Compute the changes in the cost of living figures for 2023 as compared to 2020:

Items	**Food**	**Rent**	**Clothing**	**Fuel**	**Miscellaneous**
Price (2020)	250	60	80	50	200
Price (2023)	270	80	100	50	250
% expenditure	35	20	15	10	20

In 2023, a company is looking to adjust the Dearness Allowance (D.A.) for its employees to reflect the shift in the Cost of Living Index (C.L.I.) since 2020. The employees in question were receiving a monthly wage of Rs. 200 in 2020. The task is to determine the appropriate amount of D.A. to be provided to these workers, ensuring their financial compensation aligns with the changes in the C.L.I. for the year 2023.

Solution

Let p_0 and p_n represents the price for the years 2020 and 2023 and w represents the expenditure (%).

Items	*W*	*p0*	*pn*	$I = \frac{p_0}{p_n} \times 100$	*IW*
Food	35	250	270	108	3780
Rent	20	60	80	133.33	2666.6

Clothing	15	80	100	125	1875
Fuel	10	50	60	100	1000
Miscellaneous	20	200	250	125	25000
Total	100				11821.6

$$\text{C.L.I.} = \frac{\sum IW}{\sum W} = \frac{11821.6}{100} = 118.22$$

The output indicates an increase of 18.22% in the cost of living in 2023 as compared to 2020.

However, for D.A., a person who drew 100 in 2020 at present should draw 118.22. For an employee who was drawing 200 Rs. 200 in 2020, at present he should draw

$$\frac{118.22}{100} \times 200 = 236.44$$

Hence, the D.A. should be increased by 36.44 to compensate on account of change in C.L.I.

C. Wholesale Price Index (WPI)

WPI measures the changes in the price of goods at the wholesale level. Unlike CPI, which focuses on the retail price paid by consumers, WPI tracks price changes before they reach the consumer. This index is crucial for understanding inflationary trends in the business sector.

The purpose of the Wholesale Price Index is to quantify the relative change in overall prices during a specific time frame in comparison to a predetermined base period. To examine the shifts in the general level of prices, it is necessary to gather data on the wholesale prices of marketed commodities. However, including every commodity in the index is impractical. Therefore, a representative sample reflecting the preferences, habits, and cultural norms of the populace is selected. This sample is manageable in size and encompasses commodities of a standardized quality. The index breaks these items down into five primary categories: (a) Food; (b) Fuel, Light, Power, and Lubricants; (c) Liquor and Tobacco; (d) Industrial Raw Materials; and (e) Manufactured Goods, both intermediate and finished.

Each main category is further subdivided. For instance, the ‘food’ category is split into three subcategories: (i) cereals, (ii) pulses, and (iii) others. The weighting of each group generally reflects the market value of these commodities during the base period, which should be a time of economic stability. Wholesale prices for the included commodities are gathered regularly from governmental sources and commercial hubs. Typically, price data are collected weekly, and the index number is computed accordingly. The monthly index number is derived from the Geometric Mean of these weekly indices.

In constructing this index number, the Weighted Arithmetic Mean of Price Relatives method is predominantly used. Alternatively, the Geometric Mean of Price Relatives can also be employed in the index number calculation.

Example: Compute the Wholesale Price Index Number for the 5 groups combined from the below data.

Group	Weight	Index Number
Food articles	50	241
Liquor and tobacco	2	221
Fuel, power, etc.	3	204
Industrial raw materials	16	256
Manufactured commodities	29	179

Solution

Here, we are representing the weight and Index number in W and I respectively.

Group	W	I	IW
Food articles	50	245	12250
Liquor and tobacco	2	222	444
Fuel, power, etc.	3	204	612
Industrial raw materials	16	256	4096
Manufactured commodities	29	179	5191
Total	100		22593

WPI computation

8.11.2 Quantity Index Numbers

Quantity Index Numbers, in contrast to price indices, focus on the physical volume of goods or services produced, sold, or consumed.

Industrial Production Index (IPI)

The IPI measures the output of the industrial sector, including manufacturing, mining, and utilities. It reflects the quantity of goods produced and serves as an indicator of industrial sector performance. Fluctuations in IPI can signal changes in economic growth and productivity.

The Index of Industrial Production is designed to track the relative change in industrial production levels within a country over a specified time frame, using a chosen base period for comparison. This index focuses on quantifying shifts in production volume rather than monetary value. In compiling this index, data on the production levels across various industrial sectors must be collected. Due to practical limitations, it's unfeasible to include every item of industrial production in the index. Instead, a representative yet manageable sample of items is selected. These items, of standardized quality, are categorized into six main sectors: **(i)** textile industries; **(ii)** mining industries; **(iii)** mechanical industries; **(iv)** metallurgical industries; **(v)** excise duty-bound industries; and **(vi)** miscellaneous.

Each sector is further broken down into key items. For instance, the textile sector encompasses subcategories like cotton, wool, and silk products.

For constructing these indices, the most commonly used method is the weighted arithmetic mean of production relatives. Alternatively, the weighted geometric mean of production relatives can also be employed. The weights for each sector are determined based on their relative significance, usually grounded in the net output values of the different industries during the base period.

The IPI measurement formula is as follows:

$$IPI = \frac{\sum_{i=1}^{n} q_{1i} w_i}{\sum_{i=1}^{n} w_i}$$

Here, is the quantity relative for year 1 with year 0 as the base for good i, is the weight allotted to the good i. There are n goods in the production index.

Example: Compute Industrial Production Index using weighted A.M. of production relatives based on the below data.

Items	Production in 2013	Production in 2023	Value of Net Output
A	100000	170000	350000
B	120000	160000	200000
C	5000	8000	900000

Solution

Production relative computation

Items	Production relatives
A	(170000/100000)x100=170
B	(160000/120000) x100=133.33
C	(8000/5000) x100=160

$$IPI = \frac{(170 \times 350000) + (133.33 \times 200000) + (160 \times 900000)}{350000 + 200000 + 900000} = \frac{230166000}{1450000} = 158.74$$

8.11.3 Special Purpose Index Numbers

Special Purpose Index Numbers are designed to measure specific aspects of an economy or society, often combining various data types to provide a more comprehensive overview.

A. Human Development Index (HDI)

The Human Development Index (HDI) is a composite statistic used to rank countries based on human development levels. Developed by the United Nations Development Programme (UNDP), the HDI incorporates three key dimensions:

i) **A Long and Healthy Life:** Measured by life expectancy at birth.
ii) **Education:** Measured by mean years of schooling for adults aged 25 years or more and expected years of schooling for children entering school.
iii) **Standard of Living:** Measured by Gross National Income (GNI) per capita adjusted for Purchasing Power Parity (PPP).

These dimensions are combined into a composite index between 0 and 1, where higher values indicate higher levels of human development. The HDI is widely used to assess and compare the level of development and quality of life across different countries.

There are primarily two steps to calculate the HDI.

Step 1: Developing Dimension Indices

To transform variously measured indicators into indices ranging from 0 to 1, minimum and maximum benchmarks are established. These benchmarks serve as foundational zeros and desired objectives, respectively, for the standardization of component indicators.

Goalposts for Calculating HDI

Dimension	Indicator	Minimum	Maximum
Health	Life expectancy (years)	20	85
Education	Expected years of schooling (years)	0	18
	Mean years of schooling (years)	0	15
Standard of living	GNI per capita (PPP$)	100	75000

Specifically, the justification for setting life expectancy's natural zero at 20 years comes from historical evidence showing no country in the 20th century had a life expectancy below this figure. The maximum life expectancy is positioned at 85 years, a target that has been realistic for many nations over the last 30 years. For the Gross National Income (GNI) per capita, the minimum value of $100 is chosen to account for unrecorded economic activities in the lowest-income economies, while the maximum is set at \$75,000 per capita, beyond which there is little observed gain in human development, with only a few countries currently exceeding this threshold, particularly Liechtenstein, Qatar, and Singapore.

The dimension indices are calculated using below formula:

$$\text{Dimension index} = \frac{\text{Actual value} - \text{Minimum value}}{\text{Maximum value} - \text{Minimum value}}$$

For the education dimension, the above equation is applied separately to the indicators of adult mean years of schooling and children's expected years of schooling, with the arithmetic mean of these indices subsequently calculated. This approach allows for full interchangeability between the two educational measures, a vital consideration in light of many developing countries' focus on achieving universal primary and secondary education despite lower adult educational attainment levels. Additionally, the concave nature of the transformation function from income to capabilities implies that each additional dollar has a decreasing effect on enhancing capabilities. Therefore, in income calculations, natural logarithms of actual, minimum, and maximum values are employed.

Step 2: Aggregation of Indices

The Geometric Mean (G.M.) of the three developed indices was primarily used to measure the HDI. The formula is as follows:

$$\text{HDI} = (I_{\text{health}} \cdot I_{\text{education}} \cdot I_{\text{income}})^{\frac{1}{3}}$$

The HDI value is subsequently categorized into four classes, as outlined in the 2014 HDI Report. These classes are as follows:

Class	**HDI Value**
Very high human development	0.800 and above
High human development	0.700-0.799
Medium human development	0.550-0.699
Low human development	Below 0.550

Example: Compute HDI for a hypothesized country based on the below data.

Indicator	**Value**
Life expectancy at birth (years)	66.4
Expected years of schooling (years)	11.7
Mean years of schooling (years)	4.4
Gross national income per capita (2011 PPP\$)	5150

Solution

$$\text{Health index} = \frac{66.4 - 20}{85 - 20} = 0.7138$$

$$\text{Expected years of schooling index} = \frac{11.7 - 0}{18 - 0} = 0.65$$

$$\text{Mean year of schooling index} = \frac{4.4 - 0}{15 - 0} = 0.2533$$

$$\text{Education index} = \frac{0.65 + 0.2533}{2} = 0.4517$$

$$\text{Income index} \frac{\text{In}(5150) - \text{In}(100)}{\text{In}(75000) - \text{In}(100)} = \frac{8.5467 - 4.6052}{11.2252 - 4.6052} = \frac{3.9415}{6.62} = 0.5954$$

$$\text{HDI} = (0.7138 \times 0.4517 \times 0.5954)^{\frac{1}{3}}$$

$$\text{HDI} = (0.7138 \times 0.4517 \times 0.5954)^{\frac{1}{3}} = 0.5769$$

For this hypothetical country, the calculated Human Development Index is approximately 0.58. This value suggests a relatively medium level of human development, indicating medium performance in health, education, and standard of living. The HDI is a useful tool for comparing the development levels of different countries and for assessing progress over time.

Poverty Index

The Poverty Index gauges the extent and depth of poverty in a region. It often incorporates factors like income levels, access to basic services, and inequality measures. This index is crucial for policy-making and targeting welfare programs.

8.12 Purchasing Power of Money, Deflating Using Index Number and Real Wage

It's widely acknowledged that the monetary value, or the purchasing power of money, fluctuates over time in line with price changes. When prices are consistent, the monetary value remains steady. Conversely, an increase in prices leads to a decrease in the value of money, and the opposite is true as well.

Consider a scenario where, during a reference period, the price index escalates to, for instance, 140. In such a case, the purchasing ability of one rupee diminishes to 5/7th or 100/140 of its original value in the baseline period. This implies that the monetary worth of a rupee from the baseline period has decreased to approximately 71 paise (5/7=0.71). In other words, the buying power of a rupee, in relation to a specific baseline period, is now roughly 71 paise.

This observation illustrates that the buying power of money, or its value, is essentially inversely related to the price index. The purchasing power in a given period, relative to the baseline period, can be calculated as follows:

$$\text{Purchasing power of money or value of money} = \frac{100}{\text{price index number}}$$

For instance, if the price index in 1980 is 110.3, and it was 98.4 in 1970, the 1970 rupee's purchasing power would be 110.3/98.4 of the 1980 value. Therefore, the buying power of a rupee in any given year, in terms of a 1980 rupee, can be determined by:

$$\text{Purchasing power} = \frac{\text{price index number for 1980}}{\text{price index number for the year}}$$

Moreover, the index number is an invaluable tool for converting nominal income into real income. Real income reflects the equivalent earning power in terms of the baseline year's money value. It is calculated by dividing the nominal income by a relevant Price Index, a process known as statistical deflation.

$$\text{Thus, Deflated or Real Income or wage} = \frac{\text{Money income or wage}}{\text{price index number}} \times 100$$

Consequently, the Real Wage or Income Index Number is calculated as:

$$\text{Real wage or Income Index Number} = \frac{\text{Real wage of current year}}{\text{Real wage of base year}} \times 100$$

$$= \frac{\text{Index of money wages}}{\text{Price index number}} \times 100$$

Example: Compute Income Index Number or real wages based on the below-average wages (Rs./day) of a group of workers from 2019 to 2023.

Year	2019	2020	2021	2022	2023
Average Wages	1.21	1.34	1.45	1.58	1.76
Consumer Price Index (CPI) No	96.5	102.8	101.8	102.8	110

Solution

$$\text{Income index number} = \frac{\text{Actual wages}}{\text{Price index number (CPI)}} \times 100$$

For 2019, $= \frac{1.21}{96.5} \times 100 = 1.254$

Similarly, the real wages of other years are as follows:

Year	**Average Wages**	**CPI**	**Real Wages**
2019	1.21	96.5	1.254
2020	1.34	102.8	1.304
2021	1.45	101.8	1.424
2022	1.58	102.8	1.537
2023	1.76	110	1.6

8.13 Base Shifting and Splicing of Index Numbers

Base Shifting: This method involves updating the reference point of a data series to a more recent baseline. This adjustment becomes necessary when **(a)** the original baseline has become outdated, prompting a need to select a more current period as the new baseline for enhancing the relevancy of index numbers, or **(b)** when there are multiple index number series with varying baselines that require consolidation for comparative analysis. In such instances, all series are recalibrated to align with a unified baseline.

After shifting the baseline, the recalculated index number for any given year is determined as follows:

$$= \frac{\text{Old index number for the year}}{\text{Old index number for the new base year}} \times 100$$

Example: The following table represents the wholesale price index number of a commodity with a base 2015. Shift the base to 2022.

Year	2019	2020	2021	2022	2023
WPI	113	142	162	172	187

Solution

Base Shifting from 2015 to 2022

Year	WPI (Base 2015)	WPI (Base 2022)
2019	113	(113/172)*100=65.698
2020	142	(142/172)*100=82.558
2021	162	(162/172)*100=94.186
2022	172	100
2023	187	108.721

Splicing of Index Number

Splicing is a method often employed when an index number series is initially established with a certain base year, and subsequently, for later years, a new series is created by adopting a different base year. This approach is particularly useful in scenarios involving significant changes, such as the introduction of new commodities, the exclusion of obsolete ones, alterations in quantity weights, or modifications in calculation methodologies. However, due to the distinct base periods of these series, direct comparison becomes impractical. To resolve this, a unified series with a consistent base is formed by merging the two distinct series.

Therefore, splicing refers to the process of amalgamating multiple index number series, each with its own base, into a single series unified by a common base. This unification is feasible only when there is a period that overlaps between the old and new series. The splicing technique is very similar to the base shifting process in its execution.

Example: Splice the following two series of index numbers with base (i) 2013 and (ii) 2017.

Year	2013	2014	2015	2016	2017	2018	2019	2020	2021
Series A	100	122	127	142	162				
Series B					100	112	132	152	182

Solution

Splicing of two Series of Index Numbers

Year	Old Index Number		Spliced Index Number	
	Series A	Series B	Base 2013	Base 2017
2013	100		100	(100/162)*100=61.728
2014	122		122	(122/162)*100=75.309
2015	127		127	(127/162)*100=78.395
2016	142		142	(142/162)*100=87.654

2017	162	100	162	100
2018		112	(112/100)*162=181.44	112
2019		132	(132/100)*162=213.84	132
2020		152	(152/100)*162=246.24	152
2021		182	(182/100)*162=294.84	182

8.14 Practical Applications of Index Numbers

Index numbers are not just theoretical constructs; they have practical applications in various fields. Understanding these applications is crucial for understanding the value of index numbers in real-world scenarios.

8.14.1 Inflation and Deflation Measurement

Tracking Price Changes: The most common use of index numbers is in measuring inflation and deflation. Indexes like the Consumer Price Index (CPI) and Wholesale Price Index (WPI) track changes in the price level of a basket of goods and services over time.

Monetary Policy: Central banks use these measures to formulate monetary policies. For instance, a rising CPI may prompt a central bank to increase interest rates to control inflation.

8.14.2 Economic Policy Development

Assessing Economic Health: Governments and policymakers use various index numbers, like the GDP deflator or employment indices, to gauge the health of the economy and to formulate economic policies.

Budgetary Decisions: Index numbers help in budget planning and allocation of resources by providing a clear picture of economic trends.

8.14.3 Business and Marketing Analysis

Market Analysis: Businesses use index numbers for market analysis, tracking changes in consumer prices, market demand, and purchasing trends.

Strategic Planning: These indices assist businesses in strategic decision-making, such as pricing strategies, product development, and market expansion.

8.14.4 International Comparisons

Comparing Economic Performance: Index numbers enable the comparison of economic performance between countries. For instance, indices like GDP per capita or the Human Development Index (HDI) are used to compare the economic and social well-being of different nations.

Global Trade and Investment: They are also crucial in international trade and investment decisions, helping investors and businesses understand market conditions across countries.

8.15 Challenges in Using Index Numbers

While index numbers are invaluable tools in economic and statistical analysis, their construction and interpretation come with certain challenges. Addressing these challenges is crucial for ensuring the accuracy and relevance of the index numbers.

1. Data Quality and Availability

Accuracy of Data: The reliability of index numbers is heavily dependent on the accuracy of the underlying data. Inaccurate or incomplete data can lead to misleading index values.

Availability of Data: Often, the required data for constructing an index may be unavailable or not timely, especially in less developed economies. This limitation can hinder the creation of accurate and relevant indices.

2. Choice of Methodology

Selecting the Right Method: Different methodologies can yield different results. The choice between methods like Laspeyres, Paasche, or Fisher index depends on the specific context and data availability, which can be challenging.

Dealing with Biases: Each method has inherent biases (such as substitution bias in the Laspeyres index), and choosing the most appropriate method requires careful consideration of these biases.

3. Adjusting for Seasonal Variations

Identifying Seasonal Patterns: Many economic variables exhibit seasonal patterns (like retail sales peaking during the holiday season). Failing to adjust for these variations can distort the interpretation of an index.

Applying Seasonal Adjustments: The process of adjusting for seasonal variations is complex and requires robust statistical techniques. Incorrect adjustments can lead to misinterpretation of the data.

8.16 Recent Advances and Future Trends

The field of index number construction and application is continuously evolving, driven by technological advancements and emerging needs in various sectors. Understanding these developments is crucial for staying abreast of the changes and potential future directions in this area.

1. Technological Innovations in Data Collection and Analysis

Big Data and Machine Learning: The advent of big data analytics and machine learning has revolutionized data collection and processing. These technologies enable the analysis of vast datasets, leading to more accurate and comprehensive index numbers.

Real-Time Data Processing: Technological advancements have made it possible to collect and analyze data in real-time. This capability allows for the creation of more timely and responsive index numbers, crucial in fast-paced economic environments.

Improved Survey Methods: Digital survey techniques and data collection tools have enhanced the quality and efficiency of primary data collection, which is fundamental for constructing reliable index numbers.

2. Emerging Areas of Application

Environmental Quality Indexing: There is growing interest in developing indices that measure environmental quality and sustainability. These indices, like the Air Quality Index or the Environmental Performance Index, are becoming crucial in guiding environmental policies and public awareness.

Health and Well-being Indices: In the wake of global health challenges, there is an increasing focus on indices that track health outcomes, healthcare quality, and overall well-being.

Global Connectivity Indices: With the rise of globalization, indices measuring aspects of global connectivity, such as digital adoption, trade integration, and cross-border data flow, are gaining prominence.

Social Progress Indexes: These indexes, which focus on social and cultural parameters like inclusion, freedom, and education, are becoming more refined and widely used to assess and compare social progress across different regions and countries.

The field of index numbers is dynamically expanding, with technological innovations paving the way for more sophisticated data collection and analysis methods. Additionally, the emergence of new areas of application reflects the changing priorities and challenges of our times. This evolution is not only enhancing the utility of index numbers but is also opening new avenues for research and policy-making.

8.17 Summary

This chapter has explored the multifaceted world of index numbers, highlighting their types, construction methods, applications, and challenges. From measuring inflation with the CPI to guiding policy decisions, index numbers play a crucial role in modern statistics. They distil complex economic and social data into comprehensible metrics, aiding in decision-making and analysis. While they come with challenges like data quality and methodological choices, their significance in translating raw data into actionable insights is undeniable in our data-driven world.

8.18 Self-Assessment

1. Given the following data, calculate the simple price index for 2023 using 2020 as the base year:

Item	Price in 2020 ($)	Price in 2023 ($)
A	50	55
B	20	25
C	30	35

2. Using the data from Q1, calculate the Laspeyres price index for 2023, assuming the quantities of items A, B, and C purchased in 2020 were 10, 5, and 8 units respectively.
3. For the same set of items in Q1, calculate the Paasche price index for 2023. Assume the quantities of items A, B, and C purchased in 2023 were 12, 4, and 10 units, respectively.
4. Discuss the differences in the results obtained in Exercises 2 and 3. What do these differences indicate about the Laspeyres and Paasche methods?
5. Calculate the Fisher price index for 2023 using the data from Q2 and Q3.
6. Discuss the potential biases and limitations in the CPI as a measure of inflation. Consider aspects like the choice of base year, basket of goods, and substitution bias.
7. Research and write a brief report on how the Consumer Price Index (CPI) is calculated in your country. Include the base year, basket of goods and services, and any adjustments made for seasonal variations.
8. Compute the cost of living index (CLI) based on the below data [Consider 2021 as the base year and 2023 as the current year]:

Items	Price		Weights
	2021	2023	
Food	32	48	4
Fuel	8	12	1
Clothing	16	20	3
Rent	24	18	2
Miscellaneous	25	30	1

9. Compute the chain base index numbers (2017=100) from the link relatives provided below:

Year	2017	2018	2019	2020	2021	2022	2023
Link Relatives	100	76	72	106	96	91	91

10. Using the following data prove that Fisher's ideal index number satisfies both time and factor reversal tests:

Commodity	2022		2023	
	Price	Quantity (Kg)	Price	Quantity (Kg)
A	34	52	32	52
B	30	35	25	40
C	16	55	18	50

Unit-IX

Sampling Techniques

9.1 Learning Objectives

After completing this chapter, you should be able to:

i) Understand the concept of Sample, Population, Sampling Frame, Sample Size, etc.;

ii) Differentiate between Probability and Non-Probability Sampling methods;

iii) Understand the various methods of sampling;

iv) Recognize Sampling and Non-Sampling errors;

v) Determine Sample Size.

9.2 Introduction

In this chapter, we attempt a comprehensive exploration of sampling techniques, fundamental to the field of statistics. We begin by explaining the concepts of sample, population, sampling frame, and sample size, providing a foundational understanding of sampling methodologies. Next, we probe into the distinction between probability and non-probability sampling methods, highlighting their respective applications and implications. We also cover various methods of sampling, offering insights into how researchers can effectively select samples for their studies. Additionally, we discuss the importance of recognizing and minimizing sampling and non-sampling errors, crucial for ensuring the validity and reliability of research findings. Finally, we address the critical task of determining sample size, a key aspect of research design that directly impacts the statistical power and generalizability of study results. Through this exploration, readers will gain a comprehensive understanding of sampling techniques, empowering them to make informed decisions in their research endeavours.

Researchers often opt to use samples of data due to a combination of resource constraints, time efficiency, and logistical challenges associated with collecting information from entire populations. Complete enumeration can be prohibitively expensive in terms of time, money, and manpower, making sampling a more feasible and cost-effective alternative. Logistical obstacles, such as geographical constraints or access issues, further contribute to the preference for sampling in research. Additionally, destructive testing processes, ethical considerations, and the impracticality of testing entire populations may influence the decision to use samples.

Sampling not only allows for a reduction in biases but also facilitates statistical inference, enabling researchers to make generalizations about populations based on observations from representative subsets. Ultimately, the use of data samples is a pragmatic approach, striking a balance between feasibility, efficiency, and the need for meaningful insights in various fields of research.

9.2.1 Definition and Overview of Sampling

Sampling is a crucial process that enables the testing of hypotheses regarding population characteristics. It involves the creation of a small, unbiased subset of the population for use in tests or experiments. This approach eliminates the impracticality of surveying the entire market or population, a challenge often encountered in endeavours such as statistical analysis, population surveys, and economic studies. Data sampling, as a statistical analysis technique, involves the selection, processing, and analysis of a representative subset from a population. Its application extends to identifying patterns and extrapolating trends within the broader population.

When the population is large, researchers frequently select a smaller group of elements to analyse in order to effectively address the research problem. The primary factor in this selection procedure is to ensure that the selected elements collectively reflect the characteristics of the full population, thereby ensuring the accuracy of the sample. The chosen set of elements is known as a sample, similar to extracting a cup of water from a bigger container, and the act of choosing these samples is called sampling. Every single component within the sample is designated as a sampling unit.

9.2.2 Population, Sample, Sampling Unit and Sampling Error

To understand sampling, it is important to begin with the concept of population. A *population* refers to the entire group of individuals, items, or events that share common characteristics and are the subject of study. This group is the complete set from which data is collected, and researchers aim to draw conclusions, make predictions, or generalize findings based on the analysis of this entire population. The population can be finite or infinite, depending on the scope of the study. For example, the population of all students in a particular school is finite, while the population of all possible coin toss outcomes is infinite. It is important to note that studying an entire population is often impractical, and researchers frequently employ sampling techniques to study representative subsets of populations and make inferences about the larger group.

A sample is a smaller, but representative, component or sub-set of a larger population. Statistically speaking, it comprises distinct units, elements, or observations that are chosen from the population of interest. The objective of constructing a sample is to draw conclusions about the total population by analysing the features observed within the chosen group. Sampling is a key method employed in diverse domains such as research, surveys, and data analysis. It enables researchers to swiftly and cost-effectively derive conclusions without researching the entire population.

Let us understand the concept with examples of samples - Imagine a vast assortment of ice cream flavours available at an ice cream parlour. Instead of tasting every flavour, you opt to savour a small amount from a selection of diverse flavours. The flavours you perceive are indicative of a representative subset of the complete assortment. By sampling these items, you can make a well-informed decision regarding which flavour you may choose to purchase in a larger portion. You enter a bookstore with shelves filled with different genres of books. Instead of reading every book, you decide to randomly pick a few books from each section to get an idea of the types of books available. These selected books are your sample, giving you an overview of what the entire bookstore offers. Let's say you want to find out how students in your institute feel about the cafeteria food. Instead of asking every student, which might take a long time, you randomly select a few students from each grade to share their opinions. The group of students you've chosen is your sample, and their opinions can help you make generalizations about the entire student body.

Therefore, a sample refers to a subset of individuals drawn from a larger population, representing the group from which data is collected in the research. For instance, when investigating the opinions of students within a university, one might choose to survey a sample of 100 students, though the University has a large number of students, in multiples of 100.

In the context of sampling another relevant term is *sampling unit*. Sampling units refer to the people within a population from whom measurements are collected during the sampling procedure. These units are distinct and separate things that do not overlap. It is an individual element or a group of elements that can be selected during the sampling process. It represents the basic unit from which data is collected. Extent refers to the geographical area covered by the sampling process. It defines the boundaries within which the sampling is conducted. Time represents the specific time period during which the sampling process occurs. For example, in a survey of households, the sampling units could be individual households, and within each household, the sampling units could be the male or female head of the household.. The attributes of these sample units are pivotal in defining the accuracy and precision of the sample, hence impacting the overall expenses of the sampling procedure. Furthermore, modifying the size of the sample unit might be a purposeful decision, with the goal of promoting a normal distribution in the gathered data. Thus, a sampling unit constitutes the foundational element of a dataset, representing an individual member of a population, a grouping of members, or another pre-defined unit. Its concrete definition is a fundamental prerequisite for any statistical research or study, serving as a crucial component in establishing the groundwork for data collection and analysis.

Another important concept in the context of sampling is, *sampling error*. Sampling error refers to the difference that arises from observed samples rather than a complete examination of the entire population. Significantly, as the size of the sample increases, the margin of error in the sampling decreases, emphasising the crucial role of sufficient sample sizes in reducing errors and improving the dependability of study results. Considering the above examples, the concept of sampling error manifests when the small subsets chosen for sampling fail to accurately mirror the true characteristics of the larger populations. Consider sampling ice cream flavours; if the chosen flavours represent only a fraction of the available variety, the resulting sampling error may influence your decision on a favourite flavour. Similarly, when randomly selecting books from different genres in a bookstore, a sampling error arises if the chosen books do not adequately represent the diverse range available, potentially skewing your perception of the entire bookstore's offerings. Likewise, in the context of gauging student opinions on cafeteria food, the sampling error emerges if the selected students do not authentically mirror the sentiments of the entire student body, impacting the generalizations drawn from the sampled opinions. In each case, the discrepancy between the sampled subset and the true characteristics of the larger population introduces sampling error, highlighting the importance of a representative sample for accurate conclusions.

Another relevant concept that we need to understand in this section is the concept *of standard error*. The standard error is a crucial metric in statistical analysis, determining if a discrepancy between observed and expected frequencies can be attributed to random chance. If the difference is less than 3 times the standard error, it is considered due to chance. If the difference is equal to or greater than 3 times the standard error, it is considered statistically significant. The 2 standard error (S.E.) criterion is sometimes used as a substitute for the 3 standard error criterion. The standard error is also used in significance tests and hypothesis examinations. If the discrepancy is smaller than 1.96 times the standard error, it is considered statistically insignificant, and the null hypothesis remains valid.

9.3 Significance of Sampling in Research and Data Analysis

Sampling has a huge significance in research and data analysis. Sampling serves as a strategic and practical approach that enables researchers to draw meaningful insights and make informed conclusions about populations without the need to study every individual unit. This methodological tool is pivotal in conserving resources, saving time, and overcoming logistical challenges, particularly when dealing with large or inaccessible populations. The art of sampling facilitates efficient data collection, making it a cornerstone in statistical inference, hypothesis testing, and various research methodologies. The following section gives an understanding of the significance of sampling in research and data analysis.

i) **Optimisation of Resource Utilisation:** Sampling enables researchers to collect data from a smaller portion of a population instead of examining the full group. This is especially advantageous when working with extensive populations, as it substantially diminishes the time, expense, and exertion needed for data gathering.

ii) **Time Efficiency:** Gathering statistics from the entire population can be a laborious and unfeasible task. Sampling allows researchers to expedite the acquisition of results, enabling the execution of investigations and analysis within a practical timescale.

iii) **Operational Limitations:** Occasionally, it can be arduous or unattainable to reach every individual within a population. Sampling is a pragmatic approach, enabling researchers to examine a feasible subset of the population, particularly in situations where there are geographical, organisational, or logistical limitations.

iv) **Destructive Testing:** Destructive testing refers to a method of testing that involves intentionally damaging or destroying a component or material in order to assess its performance, durability, or other characteristics. Sampling becomes essential in instances where testing is detrimental or intrusive. For instance, in medical research that involves procedures or trials that modify the condition of the person, it may be immoral or impracticable to provide the treatment to an entire community.

v) **Data Collection Feasibility:** Depending on the study's characteristics, it may be impractical to gather data from every individual within a group. Sampling allows researchers to work with a smaller and more feasible subset, ensuring that the study stays practical and attainable.

vi) **Statistical Inference :** Statistical Inference refers to the process of drawing conclusions or making predictions about a population based on a sample, using statistical techniques and methods. Sampling is crucial for drawing statistical conclusions about populations. Through the examination of a representative group, researchers can derive conclusions and formulate predictions regarding the overall population. This is a crucial element of statistical analysis and hypothesis testing.

9.4 Sampling Frame

A sampling frame is a detailed list or representation of all the elements or units that compose a population from which a sample is drawn in a research study. It is essentially a roster or inventory that serves as the basis for selecting the sample, ensuring that every member of the population has an equal chance of being included. Consider the following example, the telephone directory for a city can serve as a sampling frame when conducting a survey about telephone usage habits. Each telephone number listed represents a potential unit in the population of phone users. It provides a comprehensive list of potential participants, making it easier to randomly select phone numbers for inclusion in the sample.

Another example is a voter registration list. It is a sampling frame when conducting political polling. Each registered voter represents a potential unit in the population of eligible voters. It ensures that the sample is drawn from the relevant population, which is crucial for obtaining accurate insights into political preferences.

In an organization. an employee roster serves as a sampling frame in workforce studies. Each employee's name on the list represents a unit in the population of company employees. It enables the selection of a representative sample for studying workplace dynamics, job satisfaction, or other employment-related factors.

Thus, it can be stated that the sampling frame is the foundation of the sampling process, providing a clear list or representation of the population elements. Its importance lies in ensuring that the sample is selected from a well-defined and known set of units, allowing for the generalizability of study findings to the entire population. A robust sampling frame should include all elements of the population, ensuring that each has an equal chance of being selected. It helps in avoiding biases and inaccuracies in the sample, promoting the representativeness of the study results.

9.4.1 Types of Sampling Frame

The foundation of any sampling process lies in the careful construction of a sampling frame, a detailed inventory or representation of the elements comprising the target population. Different types of sampling frames serve as the cornerstone for selecting samples in diverse research scenarios, each offering a unique approach to ensuring a representative subset for analysis.

9.4.1.1 List-Based Sampling Frame

List-Based Sampling Frames, a fundamental type, consist of pre-existing databases or lists of individuals representing the target population. Examples include a list of registered voters, a customer database, a student roster, or a membership directory. These frames provide a structured and comprehensive basis for sample selection, ensuring that each individual in the list has an equal chance of being included in the study.

9.4.1.2 Location-Based Sampling Frame

In contrast, Location-Based Sampling Frames rely on the geographic location of the target population to compile a list. For instance, when researching people in a riverine area, the geo-location becomes the basis for the population. This type of frame is particularly useful in studies requiring representative samples from specific geographic regions, offering a spatially organized approach to sampling.

9.4.1.3 Random Digit Dialling (RDD) Sampling Frame

For telephone surveys, the Random Digit Dialling (RDD) Sampling Frame is commonly employed. It involves generating random phone numbers through a computer algorithm, accessing both listed and unlisted numbers of potential participants. This method ensures a diverse representation, crucial for the validity of telephone-based research.

9.4.1.4 Event-Based or Time-Based Sampling Frame

Certain studies define their sampling frames based on specific events, such as attendees of training sessions or webinars. This Event-Based or Time-Based approach tailors the frame to the characteristics of participants involved in particular occurrences, offering a focused sampling strategy.

9.4.1.5 Sampling Frame by Administrative Units

In specific research studies, a Sampling Frame by Administrative Units is employed, defining the frame based on administrative divisions like schools, hospitals, or government agencies. This type of frame is often utilized in studies requiring sampling within specific organizations, providing a structured approach to target units within administrative boundaries.

9.4.1.6 Membership-Based Sampling Frame

This type of sampling frame is derived from membership lists of clubs, associations, or organizations where individuals with shared interests or characteristics are members. A sampling frame is constructed from the membership list of a professional association or a community club.

9.4.1.7 Social Media-Based Sampling Frame

Utilizing social media platforms to construct a sampling frame, where individuals or accounts with specific profiles or interests are considered part of the population. Creating a sampling frame by identifying potential participants through hashtags, groups, or pages on platforms like Twitter or Facebook.

9.4.1.8 Health Records Sampling Frame

Constructing a sampling frame using health records, patient lists, or medical databases, particularly relevant in healthcare-related research. Using electronic health records to identify potential participants for a study on a specific medical condition.

9.4.1.9 Consumer Purchase History Sampling Frame

Compiling a sampling frame based on consumer purchase histories is especially useful in market research or studies related to consumer behaviour. Utilizing data from a retail store's customer purchase database to identify potential participants for a study on shopping habits.

9.4.1.10 Alumni Database Sampling Frame

Drawing a sampling frame from the database of alumni from educational institutions or organizations. Creating a sampling frame from the list of former students of a university for a study on career trajectories or educational outcomes.

9.4.1.11 Occupational Sampling Frame

Focusing on specific occupations or professions to create a sampling frame, relevant when studying groups with shared job-related characteristics. Constructing a sampling frame from a list of licensed professionals in a particular field for a study on industry trends.

Each type of sampling frame demonstrates the diverse sources and characteristics that researchers can leverage when selecting a sampling frame tailored to the specific requirements of their study. Each type of sampling frame brings its unique advantages and considerations to the research process, highlighting the importance of selecting an appropriate frame that aligns with the objectives and characteristics of the study. Each type offers a unique perspective and approach to capturing a representative sample for analysis.

9.4.2 Challenges Related to the Sampling Frame

Incomplete or outdated sampling frames can lead to underrepresentation or exclusion of certain population segments, compromising the study's validity. Maintaining the accuracy of the sampling frame over time is a challenge, especially in dynamic populations where elements may change or new elements may emerge. Some populations may not have readily available or well-defined sampling frames, posing challenges in the selection process.

i) **Incomplete or Outdated Frames:** Sampling frames may be incomplete, missing certain elements of the population, or outdated, containing information that is no longer accurate. Incomplete or outdated frames can lead to underrepresentation or exclusion of specific groups, compromising the study's external validity. Imagine a public health researcher planning a study to assess vaccination coverage in a specific urban area. The researcher decides to use a sampling frame based on the list of individuals registered at local health clinics. However, the health clinic records are incomplete, and the list has not been updated for several years.

ii) **Nonresponse Bias:** Individuals or units in the sampling frame may choose not to participate, leading to nonresponse bias. Nonresponse bias can distort the representativeness of the sample, as those who choose not to participate may have different characteristics from those who do. Consider a market research study aiming to understand consumer preferences for a new product. The researcher compiles a sampling frame from a list of registered customers who have previously made purchases from an online retail platform. However, a significant portion of these customers chooses not to participate in the survey despite multiple outreach attempts.

iii) **Frame Unavailability:** In some cases, a suitable or well-defined sampling frame may not be readily available. The absence of a reliable frame can complicate the sample selection process and may require additional efforts to construct an appropriate frame. Imagine a researcher is interested in studying the prevalence of a specific rare medical condition within a small, isolated community in a remote area. The researcher aims to collect data from individuals affected by this condition. However, there is no existing medical registry, census data, or any centralized database that provides a comprehensive list of individuals in this remote community.

iv) **Difficulty Defining Population Boundaries:** Defining the boundaries of the population accurately can be challenging, especially in dynamic or changing populations. If population boundaries are not clearly defined, there is a risk of including or excluding certain elements inadvertently, affecting the study's generalizability. Consider a social scientist conducting a study on the job satisfaction of freelance workers in the gig economy. Defining the population boundaries becomes challenging due to the dynamic and evolving nature of the gig economy, where individuals often move between traditional employment and freelance work.

v) **Cost Constraints:** Constructing or obtaining a comprehensive sampling frame can be expensive, especially for large or diverse populations. Cost constraints may limit the researcher's ability to acquire an exhaustive frame, potentially compromising the quality and inclusiveness of the sample. Think about an environmental scientist conducting a study on air quality in a densely populated urban area. The researcher intends to use a comprehensive sampling frame based on air quality monitoring stations. However, due to budget limitations, acquiring data from an exhaustive number of monitoring stations becomes financially challenging.

vi) **Sampling Frame Errors:** Errors in the construction or maintenance of the sampling frame can occur, leading to inaccuracies. Frame errors, such as duplicate entries or misclassifications, can introduce biases into the sample selection process, affecting the validity of study outcomes. For example, a survey on consumer preferences uses a mailing list as a sampling frame. Due to a technical glitch, some individuals receive multiple survey invitations, resulting in duplicate entries in the dataset. The presence of duplicate entries can skew the results, as the opinions of individuals who received multiple invitations may be overrepresented, introducing a bias into the study outcomes.

vii) **Overlapping Frames:** In some cases, multiple frames may overlap, creating challenges in selecting a unique and representative sample. Overlapping frames can lead to the inclusion of duplicate entries in the sample, distorting the accuracy of the study results. A market research study uses two different business directories as sampling frames to identify potential participants for a business survey. However, some businesses are listed in both directories, causing an overlap.

viii) **Dynamic Populations:** For populations that are constantly changing, such as transient communities or online platforms, maintaining an up-to-date sampling frame is challenging. Changes in the population may occur between the time the frame is created and the actual data collection, affecting the study's accuracy. A social media sentiment analysis study uses a sampling frame based on current platform users. However, user demographics and engagement patterns change rapidly over time, making it challenging to maintain an up-to-date frame.

9.4.3 Qualities of an Effective Sampling Frame

i) **Comprehensiveness:** An effective sampling frame must encompass every member of the target population, ensuring a holistic representation. This inclusivity minimizes bias by preventing the exclusion of relevant groups, promoting a more accurate reflection of the population under study.

ii) **Privacy and Confidentiality:** The sampling frame must adhere to ethical guidelines, addressing privacy concerns. Compliance with legal and ethical standards is essential to establishing trust and encouraging participation while safeguarding individuals' privacy.

iii) **Uniqueness:** Each element within the sampling frame should be distinct and unique, eliminating duplication or repetition of subgroups. The integration of identifiers, such as ID numbers or addresses, contributes to preserving the integrity of the sampling frame.

iv) **Accuracy:** To maintain relevance, the sampling frame should be both up-to-date and precise, offering a current depiction of the population. Regular maintenance and validation are essential for continuity sampling, preventing errors and omissions. Outdated or incomplete frames can introduce bias, impacting the generalizability of research findings.

v) **Accessibility:** An effective sampling frame should be easily accessible to researchers for sample selection. It should provide sufficient information about each unit, enabling informed decision-making during the sampling process. Accessibility fosters transparency and reproducibility in research endeavours.

vi) **Comprehensive Coverage:** The sampling frame must encompass the entire target population, including all relevant subgroups or strata. Accurate representation of diverse characteristics ensures that the sample mirrors these attributes, contributing to the achievement of a representative sample.

vii) **Reliability:** A reliable and consistent sampling frame is crucial. Ideally sourced from dependable data, it should be meticulously maintained to prevent systematic errors. Ensuring equal chances of selection for all units promotes fairness and accuracy in the sampling process.

viii) **Flexibility:** An effective sampling frame should have the ability to adapt to changes in the population structure over time. It should be dynamic enough to accommodate evolving demographics, ensuring its continued relevance and accuracy.

ix) **Representativeness:** The sampling frame should accurately represent the demographic, geographic, and other relevant characteristics of the target population. This helps in achieving a sample that reflects the diversity of the population, enhancing the external validity of the study.

x) **Cost-Effectiveness:** While maintaining quality, an effective sampling frame should be developed and managed in a cost-effective manner. Balancing the need for accuracy with resource constraints ensures efficient utilization of research funds without compromising the integrity of the sampling process.

9.5 Sample Size

The term "sample size" is a common concept in statistics and market research, especially when conducting surveys on sizable respondent populations. It pertains to the methodology employed when researching extensive population groups.

When conducting a survey on a large population of respondents, the aim is to gather insights from the entire group. However, obtaining answers or results from every individual within the population is often impractical. In such cases, researchers take a random sample of individuals that is representative of the overall population. The size of this sample holds significant importance in ensuring accurate and statistically significant results, essential for the successful execution of the study. It plays a crucial role in generalizing findings from the sample to the entire population, providing researchers with insights that can be confidently applied on a broader scale. The careful consideration and determination of an appropriate sample size enhance the reliability and validity of the study's outcomes.

Determining the appropriate sample size holds significance in both quantitative and qualitative research. For instance, in quantitative studies aiming for population generalization, a substantial and adequately large sample becomes essential. On the contrary, when employing a qualitative approach to formulate theories on organizational behaviour, a limited number of cases may suffice.

The sample size denotes the number of observations utilized to estimate characteristics of a given population, drawn from that population through the process of sampling. Making generalizations about target populations from probability samples relies on statistical probability. A larger sample size reduces the likelihood of error when extrapolating to the target population. Probability sampling, therefore, involves a trade-off between result accuracy and the resources invested in data collection, verification, and analysis.

It is important to acknowledge that the sample size taken from the population may not be universally applicable. The determination of accuracy involves considerations of confidence level and margin of error. The confidence level reflects the certainty that survey results accurately represent the target population. Increasing the sample size diminishes sampling error.

Three approaches for determining the necessary sample size will be discussed:

1. Referring to an existing pre-made sample table
2. Determining the sample size by considering the quantity of items in the scale.
3. Rough Estimate

Within this compromise, your choice of sample size is guided by:

i) The level of confidence needed in the data, indicating the certainty that collected data represents the target population's characteristics.

ii) The acceptable margin of error, signifying the precision required for any estimates derived from the sample.

iii) The planned analyses, considering the number of categories into which you intend to subdivide your data, as certain statistical techniques have minimum data case thresholds

iv) To a lesser extent, the size of the target population from which your sample is drawn.

Additionally, the level of confidence and margin of error are critical factors in calculating the accuracy of the sample size. As these considerations increase, the required sample size tends to grow, impacting the balance between research accuracy and resource investment.

9.5.1 Factors Influencing Sample Size

The assessment of sample size primarily relies on the study design and the anticipated outcome determined before initiating the study. This determination is contingent upon the statistical inference method employed, whether it be the confidence interval technique for estimation or the test of significance procedure for hypothesis testing.

In selecting the appropriate sample size, one must take into account various factors influencing the research and develop a fundamental understanding of the associated statistics. The individual can then utilize a sample size formula to integrate these considerations, enabling confident sampling with a high probability that the survey results are statistically accurate.

Various factors influence sample sizes, including the study design, sampling method, and outcome measures such as effect size, standard deviation, study power, and significance level. Disparities exist between different study designs, such as descriptive and analytical studies. Generally, descriptive studies like questionnaires and surveys necessitate larger sample sizes than analytical studies. Observational studies typically require more samples than experimental ones. The estimation also fluctuates based on grouping, control, and population size in randomized clinical trials, observational studies, and epidemiological studies. Nonrandomized clinical studies with multiple groupings require more samples, and qualitative data outcome measures necessitate larger samples than quantitative data.

Sample size calculations involve considerations of standard deviation, significance, power, and effect size. These parameters can be derived from prior literature or pilot studies. The investigator's judgment regarding the anticipated study effect significantly influences sample size estimation. Opting to detect a small effect size requires an increased sample size. The significance level (type 1 error) and study power are predetermined before the study. Typically set at 0.05 or 0.01, the significance level can be set at lower levels for enhanced accuracy, albeit at the cost of increased sample sizes. Deviating from these levels should be approached cautiously unless essential for the study design. The power is often set at a 20% chance of missing a difference or an 80% chance of detecting a statistically significant difference, ensuring a meaningful study impact.

Literature offers numerous methods for estimating sample size, and researchers should choose an effective, simple, and consistent approach. Minor errors in the formula, grouping, statistical baseline values, study design, and outcome measures can result in faulty estimation, significantly impacting the study's external validity. Considering all critical issues, especially follow-up loss, is essential, and an additional 10% of samples should be contemplated to account for potential losses during follow-up.

Balancing Precision and Practicality

Selecting the suitable sample technique requires a careful equilibrium between attaining accuracy in research results and considering practical factors such as limited resources, time limits, and logistical difficulties. This section explores the crucial elements of achieving a balance to guarantee the effectiveness of the selected sampling method.

Sampling Precision: Attaining precision in research entails acquiring outcomes that faithfully depict the attributes of the complete population. Accuracy is essential for drawing dependable conclusions and generalisations. Strategies like stratified sampling, systematic sampling, and probability sampling enhance precision by guaranteeing that the sample accurately represents different sectors of the population.

Practical factors are of equal importance to precision. The suitability of a sampling technique is heavily influenced by the availability of resources, the amount of time required, and the practicality of implementation. Researchers frequently face constraints in terms of financial resources, human resources, and time limitations. Striking a balance between accuracy and feasibility entails selecting a method that yields sufficiently precise outcomes while being compatible with the resources at hand.

Researchers must carefully balance the trade-offs and compromises that arise when considering the conflicting demands of precision and practicality. Choosing exceptionally accurate methods may require significant money and effort, rendering them impracticable in specific circumstances. On the other hand, opting for too pragmatic approaches may compromise the accuracy required for resilient and dependable results. Achieving optimal equilibrium necessitates meticulous deliberation of the study goals, accessible assets, and the desired level of precision.

Sample Size Determination: Selecting an appropriate sample size is a crucial factor in achieving a balance between accuracy and feasibility. Increased sample sizes typically enhance precision, however, they might present difficulties in terms of data gathering, analysis, and related expenses. Optimising the sample size necessitates assessing the balance between attaining the required level of accuracy and handling practical limitations.

Methodological Adaptability: Researchers are urged to embrace a versatile approach, taking into account the ever-changing character of research initiatives. Throughout the study process, it may be necessary to make alterations to the chosen sampling technique due to changes in the context, goals, and available resources. Methodological flexibility guarantees that accuracy is maintained even while adjusting to evolving practical limitations.

Effective communication and transparency are essential in explaining the reasoning behind the selected sampling technique and acknowledging any trade-offs made between accuracy and feasibility. Transparency improves the trustworthiness of the research results and enables readers to evaluate the accuracy of conclusions made from the sample.

Researchers can achieve accurate and dependable results by effectively balancing precision and practicality in their sample approaches. This ensures that the chosen methods correspond with the specific limits and needs of the study setting.

9.5.2 Statistical Techniques for Determining Sample Size

Statistical Techniques for Determining Sample Size involves methodologies crucial for determining the optimal size of a sample in research or survey studies. A carefully selected sample size is essential to ensure the reliability and representativeness of study outcomes. This field encompasses various statistical approaches that consider factors such as population size, margin of error, confidence level, and expected variance. By employing these techniques, researchers can enhance the precision and validity of their findings, contributing to the overall success.

The following are the steps to determine a sample size-

a) Define the size or number of individuals in the population
b) Specify the margin of error
c) Establish confidence level
d) Anticipate the expected variance
e) Conclude and determine sample size

a) **Define the Size or Number of Individuals in the Population:** Determining the size of the population represents a critical initial step in establishing the appropriate sample size for a study. The required sample size varies based on the true population size or the total number of individuals under consideration. Defining the population size can be challenging, particularly for complex or data-scarce populations, but it remains essential for obtaining accurate and representative results. In cases involving smaller, easily measurable groups, knowledge of the precise population size is of paramount importance. Even with larger populations, making the best possible estimate is crucial for a more accurate application of the sample size formula, ensuring reliable and reflective study outcomes.

b) **Specify the Margin of Error:** Determining the margin of error is a crucial step in research, acknowledging the inevitability of random sample errors when working with a subset of the total population. Functioning as a range, often expressed as a plus-minus (±) figure, the margin of error indicates the degree to which one is willing for the sample mean to deviate from the population mean. This range provides a level of certainty about the results. For instance, if a sample proportion of colleagues is taken with a designated 3% margin of error and reveals that 65% use voice recognition technology at home, it implies that when applied to the entire office, the actual usage may range from as low as 62% to as high as 68%. This statistical approach ensures a more nuanced and realistic interpretation of the findings.

c) **Establish Confidence Level:** Determining the confidence level is crucial for establishing certainty about the true proportion of the total population within a specific range. Commonly used confidence levels include 90%, 95%, and 99%, with researchers frequently opting for a 95% confidence level. In understanding the relationship between confidence levels and intervals (mean of error), individuals can confidently sample by combining these concepts. For example, one can assert a 95% certainty that the actual percentage of colleagues using voice recognition technology at home falls within ± three percentage points from the sample mean of 65%, ranging between 62% and 68%. The confidence level is associated with a "z-score," indicating the placement of the raw score (percentage of the confidence level) in standard deviations below or above the population mean. A z-score represents the position of the raw score (which corresponds to the percentage of confidence level) in terms of standard deviations below or above the population mean.

   Z-scores for the most common confidence intervals are:

   90% = 1.645

   95% = 1.96

   99% = 2.576

For an alternative confidence interval, refer to the z-score table to identify the accurate value for computation. In determining the sample size, one must also consider the expected variance among participant responses. Variance, in this context, refers to the degree of deviation individual sample data points exhibit from the average population. To assess the variance, researchers commonly rely on the standard deviation, which quantifies the extent of individual data points' divergence from the population average. If uncertain about the expected variance, employing a standard deviation of 0.5 ensures a sufficiently large group size for robust results.

d) **Conclude and Determine Sample Size:** The formula for estimating sample size varies according to the type of study design. However, this process can be challenging for beginners, and seeking the assistance of a statistician before initiating the study is advisable. An alternative, more straightforward method for determining sample size involves utilizing software tools. These programs consider factors such as target variance, statistical test power, and confidence interval to calculate the sample size. Notable software options include G*Power, PS, Russ Lenth's power, Minitab, Stata, R package, PASS, and the SampSize app. Unlike complex formulae, these tools simplify the calculation process. Despite the convenience offered by software, caution is essential. Users must possess a solid understanding of research design and awareness of common errors to ensure accurate results. Vigilance in controlling these errors is imperative for the reliability of the sample size determination.

Numerical Examples of Sample Size Determination

Sample Size for a Known Population

N = Population Size

Z = z-score

E = margin of error

P = standard deviation

$$\text{Sample size} = \frac{\dfrac{z^2 \times p(1-p)}{e^2}}{1+\left(\dfrac{z^2 \times p(1-p)}{e^2 N}\right)}$$

Sample size for an unknown population

Z = z-score

E = margin of error

P = standard deviation

$$\text{Sample size} = \frac{z^2 \times p(1-p)}{e^2}$$

Example: Let us try to determine the sample size from a population of 500 workers in the office so that we can determine, with a 95% confidence level and a 5% margin of error, how many workers use voice recognition technology.

Here:

The 95% confidence level translates to a z of 1.96.

The margin of error, e = 5% or 0.05

Since the standard deviation is unknown, we will set it to 0.5. So, p =0.5

The population size is known. So, N = 500.

We will calculate the sample size, n, using the first formula for a known population.

$$n = \frac{\frac{(1.96)^2 \times 0.5(1-0.5)}{0.05^2}}{1+(\frac{(1.96)^2 \times 0.5(1-0.5)}{(0.05)^2\,(500)})}$$

$$n = \frac{\frac{3.8416 \times 0.25}{0.0025}}{1+(\frac{3.8416 \times 0.25}{0.0025 \times (500)})}$$

$$n = \frac{\frac{0.9604}{0.0025}}{1+(0.954)}$$

$$n = \frac{384.16}{1.954}$$

$$n = 196.6$$

Hence the sample size is 197 respondents.

Note that if the population size were unknown, the sample size would be 385 respondents. The calculation for this is left to the reader.

9.6 Sampling Techniques

Sampling techniques play a crucial role in research, guiding the selection of a subset of a population for study. These techniques can be broadly categorized into two main types: probability and non-probability methods, each distinguished by its unique characteristics, applications, and considerations.

In probability sampling, a systematic and random selection process is employed, where each member of the population has a known and equal chance of being included in the sample. This method aims to create a representative subset that accurately mirrors the characteristics of the entire population. Probability sampling is characterized by randomization, representativeness, and its ability to allow for statistical inferences about the broader population. It is particularly suitable when precision, generalizability, and a complete sampling frame are crucial considerations. Probability sampling is the predominant method of sampling used in public opinion surveys, election polls, and other studies where the findings are intended to be generalised to a larger population. This applies regardless of the size of the overall population, whether it is as huge as a whole country or as tiny as a specific village inhabited by young girls.

Let us look at this with an example. Imagine conducting a study on the prevalence of a specific health condition within a large urban population. In this scenario, simple random sampling could be employed. Each individual in the population has an equal and known chance of being selected for the study. By randomly selecting participants, the researchers can create a sample that accurately represents the diverse health characteristics of the entire urban population. This not only allows for precise estimates of the health condition's prevalence but also enables the findings to be generalized to the broader urban community.

In another case, consider a research study aiming to explore academic performance across different grade levels in a school district. Instead of relying on simple random sampling, which might not ensure representation from each grade, the researchers opt for stratified sampling. They divide the population (students) into strata based on grade levels and then randomly select samples from each stratum. This ensures that each grade is adequately represented in the study, offering a more nuanced and accurate understanding of academic performance across the entire school district. The stratified sampling approach allows for both precision in analysing grade-specific trends and the generalizability of findings to the entire student population.

On the other hand, non-probability sampling methods do not rely on random selection, and individuals do not have a known probability of being included in the sample. Instead, selection is based on specific criteria, convenience, or the researcher's judgment. While non-probability sampling may lack the representativeness and precision of probability methods, it is often favoured in situations where time, resources, or specific subgroup emphasis are critical factors.

To understand non-probability sampling, consider a market research study aiming to gather quick insights about consumer preferences in a shopping mall. In this situation, convenience sampling could be employed. Researchers approach individuals who are conveniently accessible within the mall, such as those passing by or present in a common area. While this method lacks the randomness of probability sampling, it provides a practical and efficient way to collect initial data. The findings may not be fully representative of the entire consumer population, but convenience sampling serves the immediate need for quick insights without the extensive resources required for a more comprehensive study.

In qualitative research exploring the experiences of individuals living with a specific medical condition, purposive sampling may be chosen. Researchers deliberately select participants based on certain characteristics, such as those with diverse perspectives or varying treatment approaches. While this method lacks the randomness associated with probability sampling, it allows for a targeted and in-depth exploration of specific aspects of the phenomenon. Purposive sampling is practical when the research aims to capture rich, detailed insights from participants who possess unique perspectives, even if they do not represent the broader population uniformly.

Non-probability sampling methods are employed when flexibility, specific criteria, or resource efficiency take precedence over strict representativeness and precision. These approaches are particularly useful in the initial stages of research or when exploring specific facets of a phenomenon in depth.

The key differences between the two lies in their approach to representativeness, randomization, precision, and generalizability. Probability sampling is ideal for rigorous research requiring high accuracy, statistical analyses, and generalizability to the broader population. On the other hand, non-probability methods find their application in exploratory or preliminary studies, where practical constraints play a significant role. The choice between these methods ultimately depends on the research objectives, available resources, and the desired level of precision and generalizability for the study at hand.

Table 9.1 Differences between Probability Sampling and Non-Probability Sampling Techniques

Parameter	**Probability Sampling**	**Non-Probability Sampling**
Random Selection	Systematic and random selection of individuals from the population, ensures each member has a known and equal chance of being included.	Does not rely on random selection; individuals may be chosen based on convenience, judgment, or specific criteria.
Representativeness	Aims to create a representative subset that accurately mirrors the characteristics of the entire population.	May lack representativeness, potentially leading to biased results as the selection is not based on randomness.
Randomization	Characterized by randomization, ensuring fairness and equal opportunity for all population members.	Does not emphasize randomization, leading to a lack of fairness in selection.
Statistical Inference	Allows for statistical inferences about the broader population based on the characteristics of the sample.	Findings may lack precision and face limitations in generalizability to the entire population.
Suitability for Precision	Ideal for research requiring high accuracy, statistical analyses, and generalizability to the broader population.	More suitable for exploratory or preliminary studies where precision is not the primary focus.

Level of Control	Provides a higher level of control over the selection process, ensuring systematic and unbiased representation.	Provides less control, often relying on convenience or researcher judgment, introducing subjectivity.
Sampling Frame Requirement	Often requires a complete and updated sampling frame for accurate random selection.	Does not necessarily require a complete sampling frame, making it more flexible and suitable for situations where such a frame is unavailable.
Complexity and Resource Demand	More complex, and may demand more resources in terms of time, effort, and costs.	Less complex, and often more resource-efficient, making it suitable for scenarios with limited resources.
Generalizability	Results are more generalizable to the broader population due to the random and systematic nature of sampling.	Findings may lack generalizability, as the sample may not be representative of the entire population.
Common Applications	Commonly used in scientific research, clinical trials, and studies requiring precise statistical analysis.	Commonly used in exploratory research, qualitative studies, and situations where time and resources are limited.

9.6.1: Probability Sampling Techniques

9.6.1.1 Simple Random Sampling

Simple Random Sampling is a straightforward probability sampling method where each member of the population has an equal chance of being selected. In a simple random sample, each member of a population is allocated a unique identification, such as a number. The sample is then selected by randomly choosing individuals, sometimes with the help of an automated software programme. This method involves randomly choosing individuals from the population, ensuring that every possible combination of individuals has an equal likelihood of being part of the sample. Simple Random Sampling finds widespread application in scientific research and opinion polls. It is particularly suitable when researchers aim to create a representative subset that accurately mirrors the characteristics of the entire population. The method ensures an unbiased selection process, contributing to the generalizability of study findings. Simple random sampling is recommended for a homogeneous population, ensuring an equal chance of inclusion for all members and not influenced by previous selection. The best method is a random number table, ensuring all members have equal chances.

In a population of 1,000 students, each student is assigned a unique number. Using a random number generator, researchers randomly select 100 students to form a simple random sample. This method ensures that every student has an equal likelihood of being part of the study, making it representative of the entire student population. This sample can then be analysed to draw conclusions applicable to the broader student community.

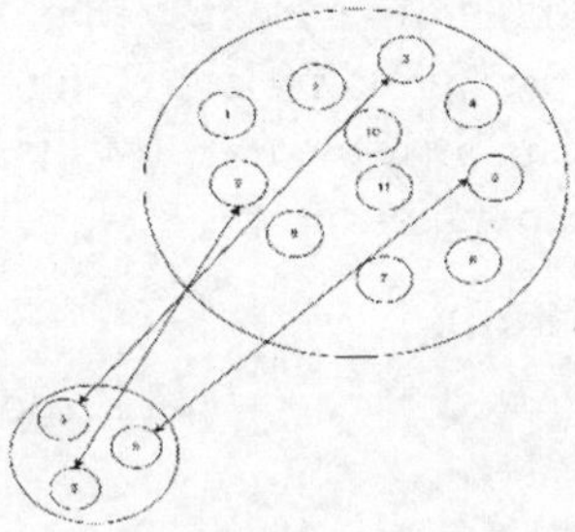

Fig 9.1: Diagrammatic Representation of Simple Random Sampling

Simple random sampling exhibits the following characteristics:

i) **Unbiased Representation:** Every individual in the population has an equal chance of being selected. This ensures that the sample is not biased towards any specific subgroup, promoting unbiased representation of the entire population.

ii) **Statistical Rigor:** Results derived from simple random sampling allow for robust statistical analyses. The method's simplicity enhances the statistical rigour of the study, enabling researchers to employ various statistical tests and make valid inferences.

iii) **Ease of Implementation:** The method is straightforward and easy to implement. Researchers can easily understand and apply simple random sampling without complex procedures, making it accessible across various research settings.

iv) **Generalizability:** The method contributes to the generalizability of study findings. Since each member has an equal chance of being included, findings from the sample are more likely to represent the characteristics of the entire population, enhancing the external validity of the study.

v) **Fair and Transparent Selection Process:** The random selection process is fair and transparent. The transparency of the method builds trust and confidence in the study, as individuals can perceive the fairness of the selection process.

vi) **Applicability to Various Study Designs:** Simple random sampling is versatile and applicable to various study designs. Whether conducting cross-sectional or longitudinal studies, simple random sampling can be adapted to suit different research objectives, adding to its flexibility.

vii) **Efficiency in Small Populations:** Particularly efficient in smaller populations. In cases where the population size is manageable, simple random sampling provides an effective and straightforward approach without the complexities associated with larger samples.

viii) **Elimination of Systematic Errors:** The method eliminates systematic errors associated with non-random sampling. By avoiding systematic biases, simple random sampling enhances the reliability of study results, providing a more accurate representation of the population.

Use of Random Numbers in Simple Random Sampling

The random numbers are a collection of digits produced by a probabilistic mechanism.

The following are the characteristics of random numbers:

1. The likelihood of any of the digits 0, 1, 2, 3, 4, 5, 6, 7, 8, or 9 appearing at any position is equal. This represents a fraction of one-tenth.
2. The presence of any two digits in any two positions is unrelated to each other. Each member of a population is allocated a distinct numerical identifier. The sample will consist of individuals from the population whose numbers match those generated from the random number table in consecutive order until the appropriate sample size is achieved.

An example of a random number table is provided in Table 9.2.

Table 9.2: Table of Random Numbers

	1	**2**	**3**	**4**	**5**	**6**	**7**
1	3158	9427	5810	2034	7641	1892	4275
2	6089	1753	3962	5402	8174	6530	2846
3	4692	7381	3210	8567	1074	5938	4025
4	2157	6873	9540	3786	5204	1469	8032
5	3621	8945	1078	6519	4306	2758	5196
6	7803	2436	6159	4892	3704	9561	1087

Example

Suppose there is a list of 60 students, and the objective is to use a simple random sampling method to select a representative sample of ten students. To execute this process, each student is initially assigned a unique number ranging from 00 to 59. The next step involves identifying ten two-digit numbers within the specified range (00–59) from a random number table. The selection process can commence from any point on the table and proceed in any direction, commonly starting from the first row and first column of Table 8.2. By focusing on the last two digits of the identified numbers, inclusion criteria are established – if a number falls within the range (00–59), it is included in the sample. Conversely, if it exceeds this range, the number is skipped, and the next one is considered in a predetermined direction. Additionally, any number that has already been selected is omitted from further consideration. Following these systematic steps ensures the unbiased and random selection of a ten-student sample from the original list of 60, promoting fairness and representativeness in the sampling process.

9.6.1.2 Stratified Random Sampling

Stratified Random Sampling is a probability sampling method that involves dividing the population into distinct strata or subgroups based on certain characteristics. Subsequently, random samples are independently selected from each stratum, ensuring representation from all relevant segments of the population. A stratified random sample is more intricate than a plain random sample. This approach involves stratifying the population into distinct subgroups, such as males and females, and then doing a simple random sampling within each subgroup. This allows for the selection of a random sample that accurately reflects the characteristics of a larger population, such as the population of a country. Stratified sampling is employed when the population exhibits heterogeneity rather than homogeneity. Stratification refers to the process of categorising individuals within a population into similar groupings before sampling. It is imperative to guarantee that every element in the population is exclusively assigned to a specific stratum. The strata must also be completely comprehensive to guarantee that no population aspect is omitted.

This method is employed when researchers want to ensure the representation of specific subgroups in the sample. It is commonly used in educational research, socio-economic studies, and any scenario where understanding variability within subgroups is crucial.

i) **Enhanced Precision:** By dividing the population into strata, the method allows for more precise analysis within each subgroup. Results are more detailed and accurate for each stratum, providing a nuanced understanding of the population.

ii) **Ensured Representation:** Each stratum is represented in the sample, addressing the risk of underrepresentation of certain groups. Ensures that findings are reflective of the diversity within the population, reducing the potential for bias.

iii) **Efficient Resource Utilization:** Resource-efficient as it focuses resources on specific strata rather than the entire population. Maximizes the efficiency of data collection, especially when some strata are smaller or less critical to the research objectives.

iv) **Increased External Validity:** Results can be more broadly generalized to the entire population. The inclusion of diverse strata enhances the external validity of the study, allowing for broader applicability of the findings.

v) **Identification of Stratum-Specific Patterns:** Facilitates the identification of patterns or trends unique to specific strata. Researchers can gain insights into variations within the population, helping tailor interventions or policies to specific groups.

vi) **Mitigation of Sampling Bias:** Reduces the risk of sampling bias associated with oversights in a simple random sampling approach. The method systematically addresses variability, ensuring a more comprehensive and accurate representation of the population.

vii) **Complexity in Implementation:** It requires a more complex implementation process compared to simple random sampling. While efficient, the method demands careful planning and understanding of the population's characteristics.

viii) **Need for Stratification Criteria:** Effective stratification requires clear criteria, which may not be applicable in all scenarios. The method's success depends on the availability and relevance of suitable stratification criteria.

Let us consider the following example, in a national health survey, the population is divided into strata based on age groups. Random samples are then independently selected from each age group, ensuring representation from various demographic segments. This allows researchers to analyse health indicators with precision within each age category, leading to a more comprehensive understanding of health trends across different age groups.

In a multinational company launching a new smartphone product, the marketing team acknowledges the significance of comprehending consumer preferences across diverse income levels. Employing stratified random sampling, potential customers are categorized into income-based strata, including low-income, middle-income, and high-income brackets. Random samples are then drawn from each stratum, ensuring the inclusion of participants from all income groups. This meticulous approach enables the marketing team to garner insights tailored to each income segment, facilitating the customization of strategies and messaging.

This method enhances targeted marketing strategies and assists in tailoring marketing strategies for specific customer segments, ensuring that promotional efforts resonate with the diverse groups within the target market. This increases precision in analysis by systematically representing each stratum. Marketers can analyse and understand the nuanced preferences and behaviours of different market segments with greater precision. This method also reduces sampling bias by providing a more accurate reflection of the overall market through the systematic representation of each income stratum. It might to some extent create complexity in implementation. Implementing stratified random sampling requires a detailed understanding of the target market's characteristics and the identification of appropriate stratification criteria, adding complexity to the process. The method can be resource-intensive, particularly when dealing with numerous strata or when collecting data from specific segments involves additional costs.

However, for further clarification, it has to be noted that, the determination of sample size within each stratum in stratified sampling involves two distinct approaches: proportional stratified sample and disproportional stratified sample. In both approaches, the fundamental principle of ensuring that every unit in the population has a chance of being selected is maintained. The discussion of these two sampling approaches follows below.

9.6.1.2.1 Proportional Stratified Sampling

Proportional stratified sampling is a method of selecting samples from different strata in a population in proportion to the size of each stratum. In this approach, the sample size from each stratum is determined based on the proportion of the total population that each stratum represents. Imagine a university with three strata: undergraduate students, graduate students, and faculty. If the total population consists of 60% undergraduates, 30% graduate students, and 10% faculty, a proportional stratified sample would select a sample from each stratum in these same proportions.

9.6.1.2.2 Disproportional Stratified Sampling

Disproportional stratified sampling is a method where the sample size from each stratum is not directly proportional to the size of the stratum in the population. This approach allows for flexibility in allocating sample sizes based on specific considerations, such as focusing more on certain strata of particular interest. In a customer satisfaction survey for a retail store with two strata - online customers and in-store customers, the management may decide to give more weight to the online customer stratum, allocating a larger sample size, even if the proportion of online customers in the total population is smaller. This could be done if the management wants more precise insights into the online customer experience.

Table 9.3: Difference between Proportional Stratified Sample and Disproportional Stratified Sample

Feature	Proportional Stratified Sample	Disproportional Stratified Sample
Basis of Allocation	The allocation of sample size is proportional to the stratum size.	Allocation of sample size is not directly proportional to stratum size.
Representativeness	Ensures that each stratum is represented proportionally in the sample.	This may result in certain strata being overrepresented or underrepresented.
Use Case	Suitable when strata sizes are relatively uniform.	Appropriate when strata sizes vary significantly and need individual consideration.
Precision in Estimates	Provides precise estimates for each.	Precision in estimates may vary, and some strata may have more accurate representation than others.
Statistical Efficiency	Generally, more statistically efficient when strata sizes are similar.	Can be less statistically efficient, especially if there is substantial variation in stratum sizes.

9.6.1.3 Systematic Sampling

Systematic sampling involves the selection of participants for a sample at regular intervals, utilizing a fixed interval or skip interval. For instance, if an interval of 5 is chosen, the sample may include the fifth, 10th, 15th, and subsequent individuals on a list. This approach is a form of probability sampling plan, providing a systematic way to choose elements from a sampling frame. The skip interval, denoted as 'k,' determines the spacing between selected elements and is calculated using the following formula:

$$\text{Skip interval (k)} = \frac{\text{Population size}}{\text{Sample size}}$$

Systematic sampling is sometimes used as a substitute for simple random sampling. It involves selecting units from a list using a skip interval called 'k.' This approach entails selecting every kth entry from the list to be included in the sample, using a randomly chosen starting point between 1 and k. For example, when the skip interval is set to 7 (k = 7) and the random start is 2, the resulting sample will consist of the 2nd, 9th, 16th, 23rd, and subsequent items from the sampling frame. If the skip interval is not an integer, it is rounded to the nearest whole number. Systematic sampling involves selecting sample units from a population at regular intervals based on time, space, or order. The selection process is a methodical procedure that consists of the following phases to determine a sample of "n" units from a population of "N" units.

Systematic sampling offers distinct advantages in research methodologies. One notable advantage is its simplicity, providing an uncomplicated and systematic approach to sample selection. The method allows for a representative subset of the population by selecting every kth element from a sampling frame, where 'k' represents the skip interval. This systematic approach ensures a structured and unbiased representation of the entire population. Additionally, systematic sampling can be more efficient than simple random sampling, especially when there is a need to cover a large population. The method strikes a balance between simplicity and representativeness, making it a valuable tool in various research contexts. Systematic sampling is a straightforward method, making it easy to understand and implement in various research contexts. It can be more efficient than simple random sampling when dealing with large populations, as it ensures systematic coverage. The systematic and structured approach eliminates the need for extensive randomization procedures, making it a practical choice in certain situations.

However, systematic sampling is not without its drawbacks. One significant disadvantage is its susceptibility to periodicity, where patterns in the population may align with the skip interval, leading to biased results. If the skip interval coincides with a periodic pattern in the population, the sample may not be truly random. Another challenge is the potential for list ordering effects, where the arrangement of elements in the sampling frame impacts the representativeness of the selected sample. Despite its efficiency, systematic sampling requires a well-organized and complete sampling frame, which may not always be readily available. Careful consideration of these advantages and disadvantages is essential when choosing systematic sampling for a research study.

Example

Assume we have a population of 100 students in a school. The list of students is arranged alphabetically, representing our sampling frame.

Desired Sample Size: Let's say we want to select a systematic sample of 10 students.

$$\text{Calculation of skip interval (k)} = \frac{\text{Population size}}{\text{Desired sample size}} = \frac{100}{10} = 10$$

This means we will select every 10th student from the list. We randomly select a starting point, let's say the 3rd student. From there, we systematically choose every 10th student. So, if our starting point is the 3rd student, the selected sample would include the 3rd, 13th, 23rd, ..., 93rd students from the list.

The following are the characteristics of a systematic sample:

i) **Regular Sampling Interval:** Systematic sampling involves selecting every kth element from a sampling frame, where 'k' is the sampling interval.

ii) **Structured Process:** The selection process follows a systematic and predetermined approach, providing a clear and organized methodology.

iii) **Random Start:** The process begins with a random start, ensuring the selection is not biased by any specific pattern in the sampling frame.

iv) **Representative Subset:** Systematic sampling aims to create a representative subset of the population by systematically choosing elements at regular intervals.

v) **Efficiency:** It can be more efficient than simple random sampling, especially when dealing with large populations, as it strikes a balance between simplicity and representativeness.

vi) **Cost-Effective:** It can be a cost-effective method, especially when compared to more complex sampling techniques, as it balances efficiency and resource use.

9.6.1.4 Cluster Sampling

Cluster sampling involves the organization of elements within a population into distinct clusters, followed by the random selection of a limited number of these clusters for further examination. Essentially, in cluster sampling, the population is partitioned into groups with diverse characteristics, termed clusters, and a subset of clusters is chosen using simple random sampling for detailed investigation. The key principle in forming clusters is to ensure that the elements within each cluster exhibit maximum heterogeneity, while the elements between clusters are as homogenous as possible. Cluster sampling involves dividing a population into distinct and different clusters, such as cities, which are then utilised as representative samples. A subset of clusters is randomly chosen from the list to participate in a study.

To achieve effective cluster sampling, the researcher must guarantee that the clusters share homogeneity based on certain unit characteristics, ensuring that each cluster represents a miniaturized version of the entire population. To ensure effective cluster sampling, the researcher carefully selects clusters based on homogeneity, using specific unit characteristics. In this approach, each chosen cluster, such as a school in a district, represents a miniaturized version of the entire population. For instance, by focusing on entire grade levels within each school as clusters, the researcher aims to capture the diversity of educational experiences across the district. This strategy ensures homogeneity within clusters, as students within the same grade level are more similar in terms of educational factors. The selected clusters act as representative subsets, reducing variability and allowing for efficient data collection from entire grade levels within schools. This method optimizes resource utilization and contributes to obtaining a diverse yet manageable sample for the research study. The assumption is that each cluster mirrors the population's diversity. Once the clusters are defined, the researcher can adopt either a one-stage, two-stage, or multistage cluster sampling approach. In one-stage cluster sampling, all elements within each chosen cluster undergo examination. Conversely, in two-stage cluster sampling, the researcher utilizes random sampling to select a subset of elements from each of the selected clusters for a more detailed analysis.

9.6.1.4.1 Types of Cluster Sampling

Single-stage Cluster Sampling involves a one-time sampling process. To illustrate, consider an organization aiming to create a sample of girls across five neighbouring towns for educational support. Employing single-stage sampling, the organization randomly selects towns (clusters) to form a sample, providing assistance to girls deprived of education in those towns.

Two-stage Cluster Sampling, in contrast, entails choosing only a few members from each cluster through systematic or simple random sampling, rather than selecting all elements of a cluster. For example, a business owner wanting to assess plant performance across various Indian locations may create clusters of plants and select random samples from these clusters for research purposes.

Taking the process further, Multiple-stage Cluster Sampling involves additional steps compared to two-stage sampling. This technique is particularly useful for intricate research spanning multiple geographies. For instance, an organization planning to survey smartphone performance across India may use multiple-stage sampling by dividing the country's population into cities (clusters). They can then select cities with the highest population and further filter them based on mobile device usage for a comprehensive analysis.

The strategy is commonly employed in statistics when it is not feasible to gather data from the full population. Therefore, it is the most cost-effective and pragmatic alternative for research statisticians. Consider the case of a researcher seeking to comprehend the patterns of smartphone utilisation in India. India's cities will cluster together in this scenario. This sampling technique is also employed in times of warfare and natural disasters to make inferences about a population when it is not feasible to collect data from every individual.

Cluster sampling exhibits the following characteristics:

i) **Variability within Clusters:** The primary principle in cluster sampling is to maximize differences or variability within each selected cluster. It is essential to ensure that the variability within clusters closely aligns with the variability present in the overall population.

ii) **Variability between Clusters:** The variability between clusters should be minimized, aiming for homogeneity within each cluster. Once clusters are chosen, data collection encompasses all units within the selected clusters.

iii) **Cost Efficiency:** Cluster sampling offers significant cost savings in data collection, as travelling costs are minimized.

iv) **Practicality in Fieldwork:** It proves practical as the researcher only needs to cover a sample of clusters, making fieldwork more manageable.

v) **Economical Data Collection:** Collecting data from nearby units is easier and more convenient, making the method cost-effective, especially for large geographic areas.

vi) **Suitability for Geographic Studies:** Cluster sampling is well-suited for research studies covering extensive geographic areas with dispersed respondents, making it a preferred method for geographical studies.

vii) **Reduced Precision:** Cluster sampling is less precise compared to sampling individual units from the entire population, as units within a cluster tend to be more homogeneous.

viii) **Decreased Efficiency with Cluster Size:** As the cluster size decreases or the number of clusters increases, the efficiency of cluster sampling is likely to decrease, impacting its effectiveness in capturing a representative cross-section of the population.

The effect of cluster size on the relative efficiency in cluster sampling is a crucial consideration in assessing the precision and effectiveness of the sampling method. Relative efficiency is a measure of how well cluster sampling performs in comparison to simple random sampling. As the cluster size increases, the relative efficiency tends to decrease. This is because larger cluster sizes contribute to increased homogeneity within clusters. When clusters are more homogeneous, the variability within each cluster becomes similar to the overall population variability. As a result, the advantage of capturing diverse characteristics within a cluster diminishes, leading to reduced efficiency. Conversely, smaller cluster sizes enhance the relative efficiency of cluster sampling. Smaller clusters are more likely to exhibit greater internal diversity, allowing for a better representation of the overall population's variability. This increased diversity within smaller clusters can result in a higher relative efficiency compared to larger clusters.

Therefore, the effect of cluster size on relative efficiency in cluster sampling highlights the trade-off between homogeneity within clusters and the precision of capturing population characteristics. Optimal cluster sizes depend on the balance between minimizing within-cluster variability and maximizing overall population representativeness.

9.6.1.5 Area Sampling

Area sampling, also known as geographic sampling or zone sampling, is a sampling method that involves dividing a population into distinct geographical areas or zones and selecting samples from these areas. This method is particularly useful when the population is spread across a large geographic region, and it may not be feasible or cost-effective to survey the entire population. Area sampling is typically used in research studies or surveys when the target population is spread across a large geographic region, and it is not feasible or practical to survey the entire population. The method is particularly suitable in situations where the population is diverse and geographically dispersed. Here are some scenarios where area sampling is often employed:

i) **Geographic Diversity:** When the population of interest is diverse and located in different geographical regions, area sampling allows researchers to capture this diversity by selecting specific geographic areas.

ii) **Cost Constraints:** Conducting a comprehensive survey of an entire population may be cost-prohibitive. Area sampling provides a more cost-effective approach by focusing on specific regions, thereby reducing the overall costs associated with data collection.

iii) **Logistical Feasibility:** In situations where it is logistically challenging to cover the entire population, area sampling provides a more manageable way to collect data. This is particularly relevant when transportation, communication, or other logistical constraints make a comprehensive survey difficult.

iv) **Heterogeneous Population:** If the population exhibits significant diversity in terms of characteristics, behaviours, or preferences across different geographic areas, area sampling allows researchers to explore this heterogeneity in a more targeted manner.

v) **Resource Optimization:** Area sampling helps optimize resources by concentrating efforts on specific regions of interest. This is especially important when the resources available for the study are limited.

vi) **Spatial Analysis:** In studies where spatial or geographic patterns are of interest, area sampling allows researchers to investigate how phenomena vary across different geographic units.

vii) **Representative Sampling:** When researchers aim to obtain a representative sample of the entire population, area sampling ensures that selected areas collectively represent the broader population in terms of relevant characteristics.

It is important to note that the decision to use area sampling should be based on a careful assessment of the research objectives, the characteristics of the population, and the available resources. While area sampling offers advantages in certain situations, researchers should also be mindful of potential biases and take steps to address them during the sampling and analysis processes.

In survey research, an area probability sample design often involves the selection of geographical areas as part of a clustered or multi-stage approach. Unlike designs where the areas themselves are the units under study, this method typically focuses on selecting specific regions, and within these chosen areas, households, individuals, businesses, or other organizations are studied and sampled. This multi-stage design allows for a systematic and structured sampling process, enabling researchers to gather data from various units within the selected geographical clusters.

Let's suppose that one block has a household size of 150, while the other block has a household size of 110. The block with a household size of 150 should have a higher likelihood of getting selected compared to the one with a size of 110. The probability in the first example should be greater than the likelihood in the second case. This principle is employed in this sampling design to assign a higher weight to clusters with larger sizes.

Example

Suppose a researcher is conducting a study on household income distribution in a large city using area sampling. The city is divided into three blocks: Block A, Block B, and Block C. The respective household sizes for each block are 200, 180, and 220. The researcher plans to select a sample of households for the study using area probability sampling.

a) Calculate the probability of selecting each block based on household size.

b) Determine the weight assigned to each block in the sampling process.

c) Discuss how the principle of assigning higher weights to larger clusters applies in this scenario.

Solution

i) Computation of Probability

$$\text{Probability of block A} = \frac{\text{Household size of block A}}{\text{Total household in all blocks}} = \frac{200}{200+180+220} = 0.33$$

$$\text{Pr obability of block B} = \frac{\text{Household size of block B}}{\text{Total household in all blocks}} = \frac{180}{200+180+220} = 0.3$$

$$\text{Pr obability of block C} = \frac{\text{Household size of block C}}{\text{Total household in all blocks}} = \frac{180}{200+180+220} = 0.36$$

ii) Determination of Weight

$$\text{Weight of block A} = \frac{1}{PA} = \frac{1}{0.33} = 3.03$$

$$\text{Weight of block B} = \frac{1}{PB} = \frac{1}{0.3} = 3.33$$

$$\text{Weight of block C} = \frac{1}{PC} = \frac{1}{0.36} = 2.77$$

The principle of assigning higher weights to larger clusters is reflected in the weights calculated. Blocks with larger household sizes have lower probabilities, resulting in higher weights. This ensures that larger clusters have a higher chance of being selected, aligning with the principle of increasing representation for larger units.

9.6.2 Non-Probability Sampling Techniques

9.6.2.1 Convenience Sampling

Convenience sampling, categorized under non-probability sampling, involves selecting units for a sample based on their easy accessibility to the researcher. The choice of units is often influenced by factors such as geographical proximity, availability at a specific time, or willingness to participate in the research. Commonly referred to as accidental sampling, this method is non-random and relies on the convenience of access. In convenience sampling, units are chosen from the population based on their ease of availability to the researcher. For instance, television channels often conduct on-the-street interviews to gauge public opinion on various issues like budget, elections, or fuel price hikes.

It's important to note that generalizing results from convenience sampling beyond the specific sample may not be appropriate. This method is most suitable for exploratory research and testing questionnaires on a pilot basis, especially in marketing studies. As the name suggests, convenience sampling is reliant on the researcher's convenience in selecting samples. This lack of a systematic approach makes it challenging to determine the representativeness of the sample, leading to potential biased estimates. Consequently, estimating sampling error, the difference between sample estimates and population parameters, is not feasible. Due to these limitations, it is advisable to avoid using convenience sampling in both descriptive and causal studies, where making definitive statements about results becomes uncertain. However, the method finds utility in exploratory designs as a foundation for generating hypotheses and in the pretest phase for activities like questionnaire testing.

i) **Accessibility and Ease:** Units are chosen based on their easy availability and accessibility to the researcher. This method is convenient for researchers as it involves selecting units with minimal effort.

ii) **Non-Random Selection:** The selection of units is not based on a random or systematic approach. Lack of randomness may result in biased samples, limiting generalizability.

iii) **Subjective Decision-Making**: The researcher subjectively decides which units to include in the sample. Sampling choices may be influenced by personal preferences, potentially impacting objectivity.

iv) **Exploratory Research Emphasis:** Often used in exploratory research to generate initial insights and hypotheses. Well-suited for preliminary investigations and hypothesis formulation.

v) **Convenience over Representativeness:** Focuses on convenience rather than ensuring a representative sample. Results may lack generalizability beyond the specific sample, limiting the study's external validity.

vi) **Useful in Pilot Studies:** Frequently employed in pilot studies, especially for testing questionnaires or study procedures. Useful in refining research instruments before implementing more rigorous sampling methods.

vii) **Quick Data Collection:** Facilitates rapid data collection due to the ease of accessing available units. Useful for studies with time constraints, but may sacrifice sample quality.

viii) **Limited Control over Variables:** Researchers have limited control over extraneous variables influencing unit selection. Increases the risk of confounding factors affecting study outcomes.

ix) **Sampling Bias Concerns:** Potential for bias as certain groups or characteristics may be overrepresented or underrepresented. Findings may not accurately reflect the broader population, raising validity concerns.

x) **Inapplicability for Definitive Statements:** Not recommended for making definitive statements about results due to inherent limitations. The lack of systematic sampling methods limits the reliability of study conclusions.

Convenience sampling, a non-probability sampling method, involves selecting units for the sample based on their accessibility to the researcher, often due to geographical proximity, availability, or willingness to participate. This approach is commonly employed in various fields, leveraging the convenience of access for data collection.

In media and journalism, street interviews conducted by television channels to gauge public opinion on pressing issues utilize convenience sampling. Individuals are selected on the street for spontaneous interviews, reflecting the convenience of their immediate availability. Retailers seeking customer feedback may opt for convenience sampling by approaching shoppers in their stores. In-store customer feedback is collected conveniently from those present at the time, providing insights into customer opinions. Convenience sampling is evident in digital marketing, especially in online surveys conducted on social media platforms. Surveys gather responses from individuals easily accessible online, allowing for quick and straightforward data collection. Convenience sampling is commonly employed in academic research, particularly on campuses. Student researchers may select participants readily available within the academic environment, such as surveying classmates or recruiting from campus events.

9.6.2.2 Purposive or Judgmental Sampling

Purposive sampling, also known as judgmental sampling, is a non-probability method that mostly depends on the researcher's discretion in selecting participants for a study. In this approach, the researcher makes his informed decisions about which elements of the population should be included in the study sample, aiming to gather valuable insights aligned with the study's objectives. The sampling design in judgmental sampling hinges on the researcher's judgment to identify individuals who can provide the most relevant and insightful information for the study's goals. This method involves focusing on participants who share similar perspectives or opinions and are willing to share their insights.

The process of employing judgmental sampling involves selecting participants based on specific criteria, such as gender, educational level, age, or socioeconomic status, that are relevant to the research objectives.

Judgmental sampling is most appropriate under certain circumstances:

a) **When the sample population is easily identified:** This method is effective when the population can be swiftly identified, and a limited amount of data is required.

b) **When the population is diverse:** Judgmental sampling is suitable when dealing with a varied population, and the researcher possesses knowledge or experience with the individuals.

c) **When there are time and resource constraints:** In situations where time and resources are limited, judgmental sampling can be a practical choice.

d) **When the research is exploratory:** Judgmental sampling is beneficial for gaining preliminary insights and understanding a topic in experimental research.

It is essential to differentiate purposive sampling, where participants are selected based on specific characteristics aligned with the study's goals, from judgmental sampling, which involves the subjective judgment of the researcher in the participant selection process. Both fall under non-random sampling strategies commonly used in qualitative research, with judgmental sampling relying on the researcher's subjective judgment to identify individuals deemed knowledgeable or reflective of a specific demographic. A researcher can choose purposive sampling when the research intends to extract significant insights from a limited sample, and the researcher has a deep understanding of the subject area, enabling him to identify crucial subjects essential for the project. This method allows to concentrate on a narrowly defined subpopulation, key individuals in a process, typical cases, or exceptional cases. Purposive sampling encompasses various techniques that employ diverse approaches to address the research question effectively.

Table 9.4: Differences between Purposive and Judgemental Sampling

Parameter	**Purposive Sampling**	**Judgmental Sampling**
Selection Method	Samples are selected based on specific criteria or purpose of the study.	Samples are selected based on the judgment or expertise of the researcher.
Bias	Introduces bias as certain types of participants are targeted.	Introduces bias as researcher's judgment may influence sample selection.
Control	Provides more control over the sample selection process.	Provides flexibility but less control over sample selection.

As a sampling method, this method holds a few advantages and disadvantages. Purposive sampling proves valuable in the exploratory research stage and helps in furnishing data that sets the groundwork for more in-depth follow-up studies. It efficiently provides this information through relatively small sample sizes, eliminating the need for a complete population list or dealing with the challenges associated with random sampling. As a non-probability method, it lacks the characteristic random selection; it tends to introduce bias into the obtained sample. As a result, the sample may not accurately represent the entire population, impeding the ability to make generalizations from the sample to the broader population.

9.6.2.3 Snowball Sampling

Snowball sampling is another non-probability sample method. In this method, the recruitment of participants in the research are done in such a way that the current research participants assist in identifying potential subjects for a study. The preferred choice is a method that minimizes risk, and justification for the chosen approach should be included in the protocol. For instance, a researcher exploring informal leadership patterns in a community might ask individuals to name others influential in that community. When the research topic is not sensitive, subjects may provide names and contact details of interested individuals. However, if the subject matter is sensitive, snowball sampling may be justifiable, with careful attention to ensuring the privacy of potential subjects. For instance, studies on networks of drug users or tracking sex partners require extreme caution in handling information obtained from one subject about another. This kind of research requires the recruitment of sample to outline steps that would minimize the risk of violating privacy. It is crucial to be careful here that current participants should not receive incentives or compensation for referrals. Acceptable alternatives to reach potential subjects involve study team members providing approved information to enrolled subjects, who can then pass it on to others who are interested or eligible. Alternatively, subjects may be asked to obtain permission from others before disclosing their contact information, ensuring the researcher doesn't directly contact potential subjects without explicit permission.

9.6.2.4 Quota Sampling

This method literally takes its meaning from the term 'quota.' In this method, the number of sample data is pre-planned in specific categories, according to some fixed quota. Quotas are pre-planned so that an appropriate representation of the population is obtained. Specific shares are allotted to categories relevant to the research being undertaken. For example, 100 women and 100 male participants in a study of literacy. It is a non-probability sampling technique where the sample of participants matches the proportions of individuals for the entire population of interest.

This type of sampling method is further divided into two types: controlled quota sampling and uncontrolled quota sampling. In the uncontrolled one, the researcher is free to choose his/her samples as per choice and convenience, but in the controlled quota sampling technique, some restrictions are imposed on the researcher that narrows down the choice of sample selection.

9.6.2.5 Quasi-Random Sampling

This technique is a type of restricted random sampling. In this method the population is arranged in some order, ascending or descending, with respect to some factors like numerically, geographically or alphabetically, and then the sample items are selected from this arranged population. For example, every nth item from the list of the population, that can be every odd or even numbered item. The first item is chosen randomly and the subsequent ones are chosen based on the regular interval as decided. This type of sampling method is applicable only where the total sample items are known or are listed. Quasi-random sampling methods are criticized for their lack of representativeness and potential for bias. However, they can be useful in certain situations where random sampling is not feasible or practical, such as in pilot studies, exploratory research, or when studying hard-to-reach populations.

9.7 Sampling and Non-Sampling Errors

Statistical error refers to the discrepancy between the values obtained from the sample (observed value) in comparison to the true value of the population. This might happen in two ways:

9.7.1 Sampling Error: This happens when the sample drawn from the population is not able to represent the population in all its perspectives. This happens as we take into consideration only a part of the population i.e., the sample, and not the whole population.

Sampling errors can be biased or unbiased in nature. Biased sampling error occurs when personal prejudice comes into existence while selecting samples from the population. Unbiased sampling errors occur out of chance. These are unintentional on the part of the researcher but arise due to chance difference in the sample items and population items. These errors can be minimised by selecting large sample sizes, using random sampling techniques and having an external review of the process.

9.7.2 Non-Sampling Error: This error happens due to reasons other than drawing a sample from the population. It encompasses the human error related to the process of the research, i.e., in properly identifying the problem, deciding upon the method and procedure to be undertaken that will be relevant to the research, etc. This also happens when the responses to the questionnaire are not accurate; the interviewer is unable to accurately conceptualize the responses from the questionnaire and the behavioural aspects that come floating during the whole process. In a study, there has been a finding that non-sampling error is a major contributing error in survey research.

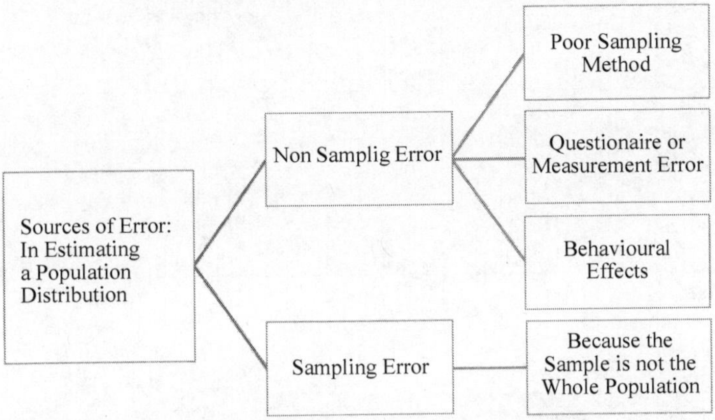

Fig 9.2: Types of Sampling Errors

The following table gives a quick recapitulation of all the sampling methods.

S. No.	Type	Used in	Probability/ Non-Probability	Selection Method	Characteristics
1	Simple Random	Various research fields	Probability	Each member has an equal chance of selection.	Equal opportunity for every unit in the population to be selected.
2	Systematic	Surveys, research	Probability	Selects every kth element from a list or population.	Simple and easy to implement, requires a starting point and sampling interval.
3	Stratified	Market research, opinion polls	Probability	Random samples taken from each stratum.	Ensures representation from all strata of the population, useful for heterogeneous populations.
4	Cluster	Census, agriculture surveys	Probability	Random clusters are selected, and all members within the chosen clusters are included.	Useful for geographically dispersed populations, reduces costs and time.

5	Convenience	Pilot studies, initial research	Non-Probability	Samples are chosen based on ease of access and availability.	Quick and inexpensive, but may not be representative of the population.
6	Judgmental/ Purposive	Qualitative research, case studies	Non-Probability	Samples are selected based on the researcher's judgment or specific criteria.	Useful for studying specific cases or phenomena, but may introduce bias.
7	Snowball	Hard-to-reach populations, network studies	Non-Probability	Initial participants recruit others from their social network.	Useful for studying hidden or marginalized populations, relies on referrals for sample expansion.
8	Quota	Market research, opinion polls	Non-Probability	Samples are selected to fulfill quotas.	Allows for stratification without random selection, but may not be representative of the population if quotas are not well-defined.

9.8 Conclusion

In this chapter, we have discussed the idea and the fundamental principles of sampling that have a huge significance in research and data analysis. It helps us zero down to insightful and informed conclusions about population without actually studying every individual unit, which helps conserve a lot of resources. We also discussed the terms like sample frame and its types and applicability with respect to different types of research requirements. Identifying and determining the sample size also plays a crucial role as it helps generalize the findings from the sample to the entire population. Thus, it has to be a comprehensive representation of the population that will enhance the reliability and validity of the study's outcome. Several sampling techniques are taken into consideration to meet this need of the study, ranging from probability sampling techniques to non-probability sampling techniques. Probability sampling techniques are used commonly in scientific research, clinical trials and in studies requiring detailed statistical analysis, the results of which are more generalizable to the broader population because each and every member of the population has got a known and equal chance of being included in the sample. Non-probability sampling techniques are used in exploratory research, qualitative studies and situations where time and resources are scarce and moreover limited to the researcher.

Though all possible means are used to make the study comprehensive and reliable, it still encounters statistical errors that can be due to mis-representation of the sample, called sampling error or due to humane behavioural aspects, called the non-sampling errors.

9.9 Summary

In this chapter, we delved into the fundamental principles of sampling, covering its definition and key concepts. We explored various probability and nonprobability sample designs, highlighting their characteristics and applications. Additionally, we discussed the importance of determining the appropriate sample size and provided practical examples to illustrate the application of these sampling designs in real-world scenarios. By the end of the chapter, readers should have gained a comprehensive understanding of sampling principles and their practical utility in research and decision-making processes.

9.10 Self-Assessment

1. What is the difference between probability and non-probability sampling techniques? Provide examples of each.
2. Explain the concept of simple random sampling. How is it different from systematic sampling?
3. Why is stratified sampling considered more representative than simple random sampling? Provide an example to illustrate your answer.
4. Describe the process of cluster sampling. What are the advantages and disadvantages of this technique?
5. What are the key characteristics of convenience sampling? When is it appropriate to use this technique?
6. How does purposive sampling differ from convenience sampling? Give an example of when each might be used.
7. Discuss the strengths and weaknesses of judgmental sampling. In what situations might this technique be most appropriate?
8. Explain the snowball sampling technique. What are the advantages and disadvantages of using this approach?
9. When would quota sampling be preferred over other sampling techniques? What are the potential pitfalls of using quota sampling?

10. How can researchers minimize bias when using non-probability sampling techniques? Provide specific strategies for reducing bias in sampling.
11. A researcher wants to study the opinions of students in a university. Which sampling method would be most appropriate for this study, and why?
12. A marketing team is conducting a survey to understand customer preferences for a new product. Which sampling technique should they use to ensure the survey results are representative of the target market?
13. A researcher is studying the impact of a new teaching method on student performance. Which sampling method would allow them to compare the performance of students before and after the implementation of the new method?
14. A company wants to survey its employees about their job satisfaction. They have a list of all employees and their departments. Which sampling method would be most effective in this scenario?
15. A researcher is conducting interviews with experts in a particular field to gather insights for a research paper. Which sampling method would be most suitable for selecting these experts?
16. A government agency is conducting a survey to estimate the unemployment rate in a country. Which sampling technique would provide the most accurate estimate of the unemployment rate?
17. A researcher wants to estimate the average height of students in a university with a margin of error of 2 cm and a confidence level of 95%. Assuming a standard deviation of 5 cm, what sample size is needed for this study?
18. A market researcher wants to estimate the proportion of customers who prefer a new brand of cereal. The researcher wants to be 95% confident that the estimate is within 3 percentage points of the true proportion. If the estimated proportion is 0.50, what sample size is needed for this study?

Unit-X

Hypothesis Testing

10.1 Learning Objectives

After reading this chapter, you will be able to:

i) Understand the concept and types of hypotheses in hypothesis testing;
ii) Differentiate between one-tailed and two-tailed tests and recognize their applications;
iii) Grasp the implications of Type I and Type II errors in hypothesis testing;
iv) Explore the significance level and its connection to confidence limits;
v) Understand the concept of degrees of freedom in hypothesis testing;
vi) Familiarize yourself with the stepwise approach to hypothesis testing for effective application.

10.2 Introduction

In this chapter, we explore the fundamental concept of hypothesis testing, a cornerstone in statistical analysis. We begin by exploring the nature of hypotheses and their types, distinguishing between one-tailed and two-tailed tests and understanding their respective applications. We then explain the intricacies of Type I and Type II errors, crucial pitfalls to avoid in hypothesis testing. Additionally, we probe into the significance level, its relationship to confidence limits, and its role in determining the strength of our conclusions. Furthermore, we explain the concept of degrees of freedom, an essential factor in hypothesis testing that affects the reliability of our results. Finally, we guide you through a stepwise approach to hypothesis testing, ensuring a comprehensive understanding and effective application of these concepts in practical scenarios. Through this journey, you will not only grasp the mechanics but also appreciate the profound impact of hypothesis testing in drawing meaningful conclusions from data in our data-driven society.

10.3 Meaning of Hypothesis Testing

Hypothesis testing serves as a statistical inference method that leverages sample data to make assertions about a population parameter or probability distribution. The process begins with the formulation of a provisional assumption concerning the parameter or distribution, known as the null hypothesis, represented by H_0. Let us understand it in more detail. Hypothesis testing is a decisive tool in research, aiming to suggest new experiments and observations. Decision-makers often test hypotheses based on available information, especially in social science where direct knowledge of population parameters is limited. This strategy allows for probability statements about population parameters. Although not absolutely proven, a hypothesis is accepted if it has withstood critical testing.

In this regard, it is important to understand the concept of hypothesis in research. By hypothesis (plural- hypotheses) we mean a proposition or a series of propositions presented as an explanation for the occurrence of a certain group of occurrences. It can be either suggested as a temporary conjecture to guide a research or acknowledged as highly plausible based on available evidence.

Table 10.1 Examples of Hypotheses from Various Fields of Study

S. No.	Field of study	Hypotheses
1.	Marketing	Increasing social media advertising spending will lead to a higher conversion rate of website visitors into customers.
2.	Human Resource	Implementing a flexible work-from-home policy will result in higher employee satisfaction and lower turnover rates.
3.	Finance	Lowering interest rates on business loans will stimulate investment and lead to increased economic growth.
4.	Operations	Streamlining the supply chain process through the implementation of Just-In-Time inventory management will reduce lead times and lower production costs.
5.	Artificial intelligence	Integrating natural language processing algorithms into customer service chatbots will improve response times and enhance customer satisfaction scores.
6.	Psychology	Increased exposure to nature will correlate with reduced levels of stress and anxiety among individuals.
7.	Sociology	The implementation of universal basic income will lead to a decrease in income inequality within communities.
8.	Economics	Raising the minimum wage will result in increased consumer spending and stimulate economic growth.

Frequently, a research hypothesis is a prognostic assertion that may be examined by scientific methodologies, linking an independent variable to a dependent one. It functions as the systematic procedure by which researchers and analysts assess assumptions, make inferences, and add to the cumulative knowledge base. Hypothesis testing is a systematic method used to confirm or disprove assertions regarding populations, parameters, and occurrences. When analysing data from a sample, a critical question arises for managers and researchers: Is the observed empirical finding merely a result of chance variation in the sample? This prompts the need for hypothesis testing, a systematic process wherein an assumption (hypothesis) about a population parameter is tested using data collected from a sample. Taking the example of a study on perfume consumption, hypothesis testing involves evaluating the likelihood that the assumed population parameter aligns with the information derived from the sample. It serves as a screening mechanism, ensuring that empirical results withstand this critical examination before further consideration. In the context of research, hypothesis testing often revolves around validating assumptions regarding the population mean. The process involves collecting data from a sample, computing the sample mean, and determining the difference between the sample mean and the hypothesized population mean. A smaller difference suggests a higher likelihood of the hypothesized population mean being correct, while a larger difference raises doubts about the validity of the assumption. Thus, hypothesis testing serves the purpose of making informed judgments about the disparities between sample statistics and hypothesized population parameters.

The process of hypothesis testing starts with the formulation of a hypothesis regarding a population parameter. The essence of the hypothesis test lies in its role as a screening question, ensuring that empirical results undergo scrutiny to assess their validity. Only after passing this crucial test do researchers dedicate substantial effort to further consideration of the empirical findings.

10.3.1 Characteristics of a Good Hypothesis

Hypotheses are crucial in research, serving as tentative explanations or predictions. While framing a hypothesis, the researcher has to be very careful that the hypothesis has some critical elements. Unless a hypothesis has those elements the value and testability as well as the results of the hypothesis testing might be confounding. They are characterized by clarity, testability, scope, specificity, simplicity, consistency, relevance etc. Understanding these attributes helps researchers develop rigorous, meaningful hypotheses that advance knowledge in their fields. Below are some of the major characteristics of a good hypothesis.

i) **Clarity and Precision:** Hypotheses should be formulated in a clear and precise manner to ensure that the inferences drawn from them are reliable and unambiguous.

ii) **Consistency with Established Facts:** Hypotheses should align with established facts and theories, ensuring consistency with existing knowledge in the field.

iii) **Relationship between Variables:** If a hypothesis is relational in nature, it should explicitly state the relationship between the variables under investigation.

iv) **Scope and Specificity:** Hypotheses should be limited in scope and specificity, focusing on a narrow aspect of the research question. Narrower hypotheses are generally more testable and yield clearer results.

v) **Simplicity:** Hypotheses should be stated in simple terms to ensure easy understanding by all stakeholders. However, simplicity should not compromise the significance or depth of the hypothesis.

vi) **Testability:** Hypotheses should be capable of being tested through empirical observation or experimentation. A hypothesis that cannot be tested hampers the progress of research.

vii) **Explanation of Phenomena:** Hypotheses should explain the observed phenomena or problem condition that prompted the need for investigation. They should provide a plausible explanation backed by empirical evidence.

viii) **Amenability to Testing:** Hypotheses should be feasible to test within a reasonable timeframe. Researchers should avoid hypotheses that require excessive time or resources for testing.

ix) **Empirical Reference:** Hypotheses should have empirical reference, meaning they should be grounded in observable data or phenomena that can be measured or observed.

x) **Falsifiability:** Hypotheses should be framed in a way that allows for the possibility of being proven false through empirical evidence. This principle, proposed by philosopher Karl Popper, is essential for scientific rigour.

xi) **Relevance:** Hypotheses should be relevant to the research question or problem statement, addressing key aspects that require investigation.

xii) **Generalizability:** Hypotheses should have the potential to generalize findings beyond the specific context of the study, contributing to broader understanding and knowledge in the field.

10.4 Importance of Hypothesis Testing in Scientific Inquiry and Decision-Making

Hypothesis testing is a crucial concept in statistics, serving as a method to ascertain whether observed phenomena or treatments have significant effects or if variables differ from each other. The primary aim of hypothesis testing is to determine if data is statistically significant and unlikely to have occurred by chance alone, essentially providing a test of significance.

One of the key benefits of hypothesis testing lies in its ability to prevent Type I and Type II errors (to be discussed in the later section). In short, Type I errors arise when the null hypothesis is wrongly rejected, while Type II errors occur when the null hypothesis is erroneously retained. By establishing a significance level and calculating the p-value, researchers can control the likelihood of these errors, enhancing the reliability of their findings. Moreover, hypothesis testing enables informed decision-making by researchers and decision-makers across various domains. For instance, in the medical field, researchers can utilize hypothesis testing to assess the efficacy of new drugs, while business analysts can employ it to evaluate the performance of marketing campaigns. By relying on evidence obtained through hypothesis testing, decision-makers can make informed choices, making better decisions based solely on speculation or conjecture. Hypothesis testing is extensively utilized to examine strategies and facilitate data-driven decisions. For instance, business owners can employ hypothesis testing to gauge the potential profitability of new products and through the examples shown above, the relevance of such scientific queries can be answered.

10.5 Types of Hypotheses

10.5.1 Null Hypothesis and Alternative Hypothesis

In statistical analysis, the concepts of null hypothesis and alternative hypothesis play pivotal roles. The null hypothesis is established when there is an assumption that two methods, such as Method A and Method B, exhibit equal effectiveness or superiority. Conversely, the alternative hypothesis emerges when there is a belief that one method surpasses the other. Symbolically, the null hypothesis is denoted as H_0, while the alternative hypothesis is represented as H_a. For instance, in testing the hypothesis that the population mean (μ) equals a hypothesized value of 100, the null hypothesis would state that the population mean is indeed equal to 100, expressed as H_0: $\mu = 100$.

Let us consider an example in the context of a manufacturing company evaluating the effectiveness of two different production methods, Method A and Method B, in terms of producing identical products.

Null Hypothesis (H_0): There is no difference in the mean production output between Method A and Method B. Symbolically, it can be expressed as H_0: $\mu_A = \mu_B$, where μ_A represents the mean production output of Method A and μ_B represents the mean production output of Method B.

Therefore, the null hypothesis (H_0) is a foundational statement regarding a population parameter, like the population mean, which is presumed to be accurate. It serves as the initial premise from which we assess whether the specified value in the null hypothesis is probable.

Using the examples from the above table, let us see how we can explore null and alternate hypotheses.

The first five propositions from the five various fields can be represented as in Table 10.2:

If our sample results do not align with the null hypothesis, we must consider an alternative explanation. The conclusion we draw when rejecting the null hypothesis is known as the alternative hypothesis because it is an alternate to the previously proposed null hypothesis. In simpler terms, the alternative hypothesis presents alternative scenarios to the null hypothesis.

Alternative Hypothesis (Ha): Method A has a higher mean production output than Method B. Symbolically, it can be expressed as Ha: $\mu_A > \mu_B$.

To test these hypotheses, the company may collect production data from both methods over a specified time period and perform statistical analysis, such as a t-test or ANOVA, to determine if there is a statistically significant difference in the mean production outputs.

If the statistical analysis yields a p-value less than the chosen significance level (usually 0.05), the null hypothesis would be rejected in favour of the alternative hypothesis, indicating that Method A is indeed superior to Method B in terms of production output. Otherwise, if the p-value is greater than 0.05, there would be insufficient evidence to reject the null hypothesis, suggesting that Method A and Method B have similar production outputs.

If we accept H_0, we reject H_a, and conversely, if we reject H_0, we accept H_a. For H_0: $\mu = 100$, we can consider three potential alternative hypotheses as follows: the population mean differs from 100 (i.e., it may be greater or less than 100). Additionally, the alternative hypothesis posits that the population mean is greater than 100, and alternatively, that the population mean is less than 100.

Table 10.2 Examples of Null and Alternative Hypotheses

H_0 **(Null Hypotheses)**	H_a **(Alternative Hypothesis)**
Increasing social media advertising spending will not lead to a higher conversion rate of website visitors into customers.	Increasing social media advertising spending will result in a higher conversion rate of website visitors into customers.
Implementing a flexible work-from-home policy will not result in higher employee satisfaction and lower turnover rates.	Implementing a flexible work-from-home policy will lead to higher employee satisfaction and lower turnover rates.

Lowering interest rates on business loans will not stimulate investment and will not lead to increased economic growth.	Lowering interest rates on business loans will stimulate investment and lead to increased economic growth.
Streamlining the supply chain process through the implementation of Just-In-Time inventory management will not reduce lead times and will not lower production costs.	Streamlining the supply chain process through the implementation of Just-In-Time inventory management will reduce lead times and lower production costs.
Integrating natural language processing algorithms into customer service chatbots will not improve response times and will not enhance customer satisfaction scores.	Integrating natural language processing algorithms into customer service chatbots will improve response times and enhance customer satisfaction scores.
There is no significant difference between the average monthly sales revenue before and after implementing the new marketing strategy.	There is a significant difference between the average monthly sales revenue before and after implementing the new marketing strategy.
There is no significant difference between the job satisfaction levels of employees in the morning shift and those in the evening shift.	There is a significant difference between the job satisfaction levels of employees in the morning shift and those in the evening shift.
There is no significant difference between the mean exam scores of students who attended traditional lectures and those who participated in online classes.	There is a significant difference between the mean exam scores of students who attended traditional lectures and those who participated in online classes.

The null hypothesis and alternative hypothesis are established before collecting data. The null hypothesis is the hypothesis being examined for rejection, while the alternative hypothesis includes all possible outcomes. The significance level (α) is set to minimize the likelihood of incorrectly rejecting the null hypothesis. The null hypothesis must be precise and specific. The null hypothesis is designated as such when rejecting it, despite being true, holds significant consequences. This is due to setting a very low significance level (α), representing the probability of erroneously rejecting the null hypothesis when it's true.

10.5.2 Levels of Significance

Hypothesis testing is a crucial concept, typically with a significance level set at a certain percentage chosen by the researcher. The level of significance in hypothesis testing is the critical point at which researchers are prepared to reject the null hypothesis, indicating the highest likelihood of committing a Type I error. A 5% significance threshold indicates a 5% tolerance for erroneous rejection. This percentage indicates that the null hypothesis (H_0) will be rejected if the observed evidence has less than a 0.05 probability of occurring under H_0. In essence, a 5% significance level implies a willingness to accept a 5% risk of rejecting H_0 when it's true. This significance level is predetermined before hypothesis testing to establish decision criteria. Similar to jurors setting criteria in a criminal trial, researchers collect data to disprove the null hypothesis based on the likelihood of selecting a sample mean from a population, typically set at 5% in behavioural research. When the probability of obtaining a sample mean is less than 5% under H0, the null hypothesis is rejected. Choosing the appropriate level balances the likelihood of committing a Type I mistake and the need to identify genuine effects or correlations in the data. The level of significance is crucial in hypothesis testing as it helps researchers make judgments based on evidence.

The significance level, typically denoted as α (alpha), represents the probability threshold used to determine whether to reject the null hypothesis. A significance level of 0.05 means that there is a 5% chance of incorrectly rejecting the null hypothesis when it is actually true. Similarly, a significance level of 0.01 means there is a 1% chance of making this error.

To assess the statistical significance of a result, the researcher must initially compute the p-value, representing the probability of observing an effect under the assumption that the null hypothesis is true. If the calculated p-value is lower than the predetermined significance level (α), typically 0.05, the null hypothesis is dismissed. Conversely, if the observed p-value exceeds α, theoretically, the null hypothesis is upheld. However, researchers often increase the sample size to ascertain if the significance level is attained. Generally, the interpretation of the p-value based on a 10% significance level is as follows:

- If $p > 0.1$, no assumption is made regarding the null hypothesis.
- If $0.05 < p \leq 0.1$, there is a slight inclination towards rejecting the null hypothesis.
- If $0.01 < p \leq 0.05$, a considerable inclination towards rejecting the null hypothesis is noted.
- If $p \leq 0.01$, a strong inclination towards rejecting the null hypothesis is evident.

For instance, if the calculated p-value is 0.03, indicating a 3% chance of observing a larger difference in our research under the null hypothesis, we assess its statistical significance. Given that the chance is below 5%, we reject the null hypothesis and accept the alternative hypothesis.

For instance, consider a scenario where a researcher conducts a study to evaluate the effectiveness of a new medication for treating a specific ailment, using a significance level of 0.05. If the study yields a p-value of 0.03, which falls below the 0.05 threshold, the researcher would reject the null hypothesis. This would lead to the conclusion that there is adequate evidence to support the alternative hypothesis, indicating that the drug is effective.

Conversely, if the researcher had chosen a significance level of 0.01 and obtained the same p-value of 0.03, they would not be able to reject the null hypothesis due to the p-value being above 0.01. In both instances, the decision to support or reject the null hypothesis is influenced by the predetermined significance level and the strength of evidence in the data. However, the stricter significance level of 0.01 reduces the likelihood of committing Type I errors, thereby adopting a more conservative approach.

10.6 One-Tailed & Two-Tailed Tests

Let us consider an example where a business analyst is testing the effectiveness of a new marketing campaign aimed at increasing traffic to the company's website. The analyst wants to test whether there is a significant difference in the average daily website visits before and after the campaign. There are two outcomes that the analyst may want to test. In the first, the analyst may want to examine if the average daily visits to the company's website is significantly greater after the campaign than before. In the other outcome, the analyst may want to examine if there is a significant difference in the average daily visits before and after the campaign. In the second case, the analyst is considering both increases as well as decreases in the number of average daily visits.

These two outcomes lead to two different tests of the hypotheses that we can generate. In the first case, the null hypothesis (H_0) is that the number of average daily visits before the campaign is less than or equal to the number of average daily visits after the campaign. The alternate hypothesis (H_A) states that the number of average daily visits after the campaign is significantly greater than the number of average daily visits before the campaign. This leads to the consideration of a one-tailed or directional test.

The second outcome lets us state the null hypothesis (H_0) that the number of average daily visits after the campaign is equal to the number of average daily visits before the campaign. The alternate hypothesis (H_A) is that the number of average daily visits after the campaign is significantly more or significantly less than before the campaign. This leads to the consideration of a two-tailed or non-directional test.

In a two-tailed test, the probability of an unlikely outcome is placed into the two tails of the distribution (hence the name), while in a one-tailed test, the region of rejection is placed in the tail specified by the alternate hypothesis. When uncertainty exists regarding the directionality of the difference being tested, a two-tailed test is preferred, ensuring that extreme values on both ends of the distribution are considered. On the other hand, if the research hypothesis implies a specific direction of difference, a one-tailed test is more suitable, focusing on only one tail of the distribution.

Let μ_{before} and μ_{after} represent the number of average daily visits before and after the campaign, and the level of significance desired is 5 per cent or 0.05. Then we can write the null and alternate hypotheses for the two outcomes as follows:

One-tailed test:

H_0: $\mu_{after} \leq \mu_{before}$

H_A: $\mu_{after} > \mu_{before}$

Two-tailed test:

H_0: $\mu_{after} = \mu_{before}$

H_A: $\mu_{after} \neq \mu_{before}$

We can then depict the two kinds of tests in the diagram below:

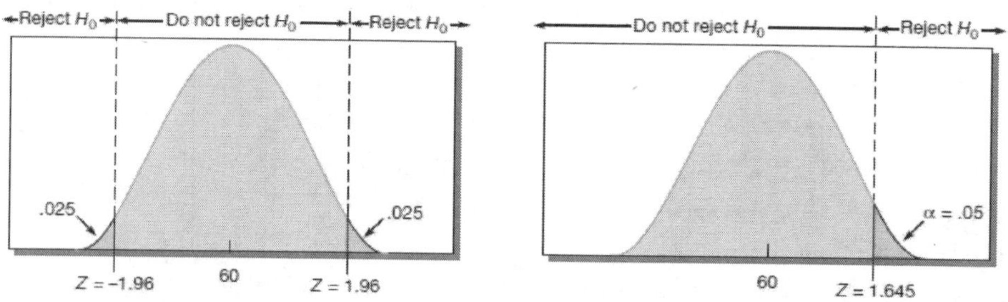

Figure 10.1 Two-tailed and One-tailed Tests at the 5 Per Cent Level of Significance

We may then use a standard T-test to see if the null hypothesis can be rejected or cannot be rejected. If the null hypothesis can be rejected, we may say that the alternate hypothesis may be accepted. However, if the null hypothesis cannot be rejected, then we do not have the evidence to accept the alternate hypothesis.

10.7 Types of Error

Since our tests of hypotheses are based on samples and not the entire population, and using probability theory, we cannot be certain if an error has been made by our tests. Using samples leads to two kinds of errors: Type I and Type II. A Type I error occurs when we reject the null hypothesis when, in fact, the null hypothesis is true. On the other hand, a Type II error occurs when we "accept" the null hypothesis (or fail to reject the null hypothesis) when the null hypothesis is not true. In simple terms, a Type I error means rejecting hypotheses that should have been accepted, while a Type II error implies that we accept a hypothesis that should be rejected.

The following table summarizes the kinds of errors.

Table 10.3 Type I and Type II Errors

Actual State in the Population	**Decision**	
	Accept H_0(fail to reject H_0)	**Reject H_0**
H_0 is true	No error (correct decision)	Type I error
H_0 is not true (false)	Type II error	No error (correct decision)

Thus, we see that when an analyst or researcher is choosing a significance level, they are also picking the tolerance for a Type I error. The significance level picked is usually called α (alpha). A Type II error is the probability of failing to reject the null hypothesis when the null hypothesis is not true, and often labelled as β (beta).

The researcher or analyst cannot reduce Type I and Type II errors without increasing the sample size. This is because Type I and Type II errors are inversely related. Thus, when we reduce the probability of a Type II error, we increase the probability of a Type I error, and vice versa. In different scenarios, different types of errors may be more serious, and we may have to make a trade-off by examining the costs and losses associated with both types of errors. Typically, in marketing situations, Type I errors are considered to be worse than Type II errors. In a pharmaceutical situation, Type II errors will mean that we may poison a set of customers of a medicine that we have not rejected as poisonous, but is actually poisonous. These kinds of differences may lead to different values of α and β being adopted in different situations.

10.8 Degrees of Freedom

Degree of freedom (df) is a concept used in statistics, particularly in the context of hypothesis testing and the analysis of variance (ANOVA). It represents the number of values in the final calculation of a statistic that are free to vary. In simpler terms, it is the number of independent pieces of information available to estimate another piece of information.

Degrees of freedom represent the number of independent values that a statistical analysis can estimate or the number of values that are free to vary as parameters are estimated. It's often denoted by v or df and is calculated as the sample size minus the number of restrictions.

For example, if you randomly sample 10 American adults to measure their daily vitamin intake and use a one-sample t-test to determine if the mean intake is equal to the recommended 1000 mg, the test statistic t has 9 degrees of freedom ($df=n-1=10-1=9$).

In reporting results, degrees of freedom are often noted alongside the test statistic. For instance, if your calculated t value is 1.41 with a corresponding p-value of 0.19, you would report: "The participants' mean daily calcium intake did not differ from the recommended amount of 1000 mg, t(9) = 1.41, p = 0.19. In hypothesis testing, degrees of freedom are important because they determine the distribution of test statistics. For example, in a t-test, the degrees of freedom are related to the sample size and affect the shape of the t-distribution. In ANOVA, degrees of freedom are used to calculate the variability between groups (the numerator degrees of freedom) and the variability within groups (the denominator degrees of freedom). These values are crucial for determining whether the differences between group means are statistically significant. In general, degrees of freedom are a way to account for the variability in a dataset and are crucial for making accurate statistical inferences.

10.8.1 Uses of Degrees of Freedom

i) **Estimation of Population Parameters:** Degrees of freedom helps us to estimate population parameters from sample data. By accounting for the number of values that are free to vary, degrees of freedom help ensure that our estimates are not biased.

ii) **Hypothesis Testing:** In hypothesis testing, df is crucial for determining the appropriate distribution of test statistics. They help us calculate p-values and make decisions about the significance of our results.

iii) **Model Complexity:** Degrees of freedom also reflect the complexity of the statistical model being used. More complex models often require more degrees of freedom to estimate parameters accurately.

iv) **Control of Type I Errors:** Degrees of freedom play a role in controlling Type I errors (false positives) in statistical analyses. By correctly accounting for the variability in the data, degrees of freedom help ensure that our conclusions are valid.

10.9 Confidence Limits

Confidence limits, also known as confidence intervals, are a range of values calculated from sample data that is likely to contain the true population parameter with a certain level of confidence. In other words, confidence limits are the lower and upper bounds of a range of values that estimate the mean of a normal population. When we estimate a population parameter (like a mean or a proportion) from a sample, we often want to quantify the uncertainty around that estimate. Confidence limits provide a range of values within which we can reasonably expect the population parameter to fall. The confidence level (e.g., 95% confidence) indicates the probability that the interval will contain the true parameter if we were to take many samples and compute intervals for each sample.

10.9.1 Uses of Confidence Limits

i) **Estimation:** Confidence limits provide a range of plausible values for a population parameter based on a sample. For example, if we estimate the average height of a population to be 170 cm with a 95% confidence interval of [165 cm, 175 cm], we are saying that we are 95% confident that the true average height falls within this range.

ii) **Hypothesis Testing:** Confidence intervals can also be used for hypothesis testing. If the null hypothesis predicts a specific value for a parameter and this value falls outside the confidence interval, we may reject the null hypothesis.

iii) **Comparison of Groups:** Confidence intervals can be used to compare groups or conditions. If the confidence intervals of the two groups do not overlap, it suggests that there is a significant difference between the groups.

iv) **Decision Making:** Confidence intervals help decision-makers assess the reliability and precision of estimates. For example, in medical research, confidence intervals around the effectiveness of a treatment help doctors make informed decisions about its use.

10.10 Steps in Hypotheses Testing

Hypothesis testing is a method used in statistics to evaluate a hypothesis about a population parameter based on sample data. When researchers interpret their findings, they must determine if the results could have occurred randomly. Hypothesis testing is a methodical approach to determining whether the outcomes of a study support a specific theory that applies to a larger population. Using sample data, hypothesis testing evaluates a hypothesis regarding the population. It gauges the likelihood of the observed result being due to chance or if it is too significant to be attributed to chance alone.

Here are the detailed steps involved in hypothesis testing:

Step 1: Formulate the Null Hypothesis (H_0)

The null hypothesis is the default assumption or the statement being tested. It typically states that there is no effect, no difference, or no relationship in the population. Formulating the null hypothesis (H_0) is a fundamental step in hypothesis testing. It acts as the basis for comparison with the alternative hypothesis (Ha), which suggests the presence of an effect or relationship. The null hypothesis is denoted as H_0 and is crucial in guiding the statistical analysis by providing a reference point for evaluating the evidence from the sample data. It should be clear, specific, and testable based on the available data, helping researchers draw meaningful conclusions about the population parameter being tested. The initial stage requires formally articulating the null hypothesis (H_0). It is crucial to define these hypotheses clearly, taking into account the research problem's nature.

Examples

i) Ms. Patel from the Mechanical Engineering Department wants to test the load-bearing capacity of an old bridge, which she believes should exceed 10 tons. In this case, she can express her hypotheses as follows:

   Null hypothesis H_0: $\mu = 10$ tons

ii) There is no difference in sales performance between customers who receive a discount coupon and those who do not.

   Null Hypothesis H_0: $\boldsymbol{\mu 1} = \mu 2$, where $\mu 1$ represents the mean sales performance of customers who receive a discount coupon and $\mu 2$ represents the mean sales performance of customers who do not receive a discount coupon.

Step 2: Formulate the Alternative Hypothesis (H_a)

The alternative hypothesis is the opposite of the null hypothesis. It states what we are trying to find evidence for and is denoted as H_a. Formulating the alternative hypothesis (H_a) is a crucial step in hypothesis testing, as it specifies the researcher's belief about the population parameter being tested. The alternative hypothesis represents the opposite of the null hypothesis (H_0) and suggests that there is a relationship, effect, or difference in the population. Depending on the nature of the hypothesis, the alternative hypothesis can be two-tailed, indicating a difference without specifying the direction; one-tailed, specifying the direction of the difference; or non-directional, indicating a difference without specifying the direction. The alternative hypothesis guides the statistical analysis and interpretation of results, helping researchers draw meaningful conclusions about the phenomenon under study.

For the same cases discussed above, the alternative hypothesis (Ha) will be,

i) For Ms. Patel's hypothesis regarding the load-bearing capacity of the old bridge, the alternative hypothesis (H_a) can be formulated as follows:

   The load-bearing capacity of the old bridge exceeds 10 tons. This alternative hypothesis suggests that there is a difference in the load-bearing capacity of the old bridge compared to the specified value of 10 tons in the null hypothesis.

   Alternate Hypothesis H_a: $\mu > 10$

ii) For the second one, the alternative hypothesis is, 'there is a difference in sales performance between customers who receive a discount coupon and those who do not'.

$H_{a:\mu} 1>\mu 2$ or $:1>2$

$H_{a:\mu} 1<?\mu 2$ or $:1<2$

$H_{a:\mu} 1\neq\neq\mu 2$, depending on the direction of the expected difference in sales performance.

Step 3: Choose the Significance Level (α)

As discussed in an earlier section, the significance level, denoted as α (alpha), is the probability of rejecting the null hypothesis when it is actually true. Commonly used values for **α are** 0.05 (5%) or 0.01 (1%), but the choice depends on the specific context of the study. The significance level (α) plays a crucial role in determining the likelihood of making a Type I error, which occurs when the null hypothesis is incorrectly rejected. A common choice for α is 0.05 (5%), which means that there is a 5% chance of rejecting the null hypothesis when it is actually true. However, researchers can also choose a lower significance level, such as 0.01 (1%), for more stringent criteria, reducing the risk of Type I errors but potentially increasing the risk of Type II errors (failing to reject the null hypothesis when it is false) if the sample size is small or the effect size is small.

For example, consider a marketing scenario where a researcher wants to test if a new advertising campaign leads to a significant increase in sales. The null hypothesis (H_0) could be that there is no difference in sales between the new and old advertising campaigns. The alternative hypothesis (H_a) would then be that there is a significant difference in sales between the two campaigns.

By choosing a significance level of α = 0.05, the researcher is saying that they are willing to accept a 5% chance of incorrectly rejecting the null hypothesis, assuming that there is no real difference in sales between the two campaigns. If the p-value of the hypothesis test is less than 0.05, the researcher would reject the null hypothesis and conclude that there is sufficient evidence to support the alternative hypothesis that the new advertising campaign leads to a significant increase in sales. However, if the p-value is greater than 0.05, the researcher would fail to reject the null hypothesis and conclude that there is not enough evidence to support the alternative hypothesis.

Step 4: Select the Test Statistic

The choice of test statistic depends on the hypothesis being tested, the type of data, and the sample size. Choosing the correct test statistic is paramount in hypothesis testing as it ensures the analysis is suitable for the data and research question at hand. Using an inappropriate test statistic can lead to inaccurate results and flawed conclusions. The right choice guarantees the validity of the analysis, ensuring that the results accurately reflect the underlying relationship or effect being studied. Additionally, it enhances the reliability of the results, making them more replicable and generalizable. Correctly choosing the test statistic also improves the interpretability of the results, facilitating a clearer understanding of their implications. Furthermore, it enhances the efficiency of the analysis by providing more precise estimates and reducing the risk of errors. Lastly, using the correct test statistic is ethically important as it ensures that decisions made based on the results are well-founded and not misleading. A few things that impact the researchers' choice of test statistic are:

i) Type of Data: Sales data is likely to be continuous and may follow a normal distribution, especially if the sample size is large. However, if the sample size is small or the population standard deviation is unknown, a t-statistic would be more appropriate than any other test.

ii) Hypothesis Being Tested: The hypothesis is comparing the mean sales between two groups (old and new advertising campaigns), which is a classic scenario for a t-test.

iii) Sample Size: If the sample size is large (typically more than 30), a z-test could also be used as an approximation of the t-test.

Table 10.4 Statistical Tests for Hypothesis Testing in Various Scenarios

Hypothesis Test	Type of Analysis	Test Statistic	Remarks
Mean Difference (Two Independent Samples)	Two-Tailed Test	t-statistic	Commonly used for comparing the means of two groups with unknown population standard deviations.
Mean Difference (Two Independent Samples)	One-Tailed Test (Directional)	t-statistic	Used when the hypothesis specifies the direction of the difference in means.
Mean Difference (Two Independent Samples)	Two-Tailed Test	z-score	Appropriate when the sample size is large (typically >30) and population standard deviations are known.
Mean Difference (Two Independent Samples)	Chi-Square Test	Chi-square statistic	Used for categorical data, such as comparing proportions between two groups.
Variance Difference (Two Independent Samples)	F-Test	F-statistic	Used to compare the variances of two groups.

Mean Difference (Paired Samples)	Two-Tailed Test	t-statistic	Used when comparing the means of two related groups (e.g., before and after intervention).
Correlation	Two-Tailed Test	Pearson correlation coefficient (r)	Used to measure the strength and direction of a linear relationship between two continuous variables.
Difference in Proportions (Two Independent Samples)	Two-Tailed Test	z-score	Used to compare proportions between two groups.
Association between Categorical Variables	Chi-Square Test	Chi-square statistic	Used to test the independence of two categorical variables.
Difference in Medians (Non-Normal Data)	Mann-Whitney U Test	U-statistic	Non-parametric test for comparing medians of two independent samples.
Difference in Medians (Paired Samples, Non-Normal Data)	Wilcoxon Signed-Rank Test	W-statistic	Non-parametric test for comparing medians of two related samples.
Association between Ranked Variables	Spearman's Rank-Order Correlation	Spearman's rho (ρ)	Non-parametric measure of association between two ranked variables.
Comparison of Means (Three or More Independent Samp	ANOVA	F-statistic	Used to compare means of three or more groups.

Step 5: Determine the Critical Region

The critical region is the set of values of the test statistic that leads to the rejection of the null hypothesis. It is determined based on the chosen significance level and the distribution of the test statistic. Determining the critical region is a crucial step in hypothesis testing, as it defines the boundaries beyond which the null hypothesis will be rejected. The critical region is determined based on the chosen significance level (α) and the distribution of the test statistic.

For example, in a z-test for a population mean, if the significance level is $\alpha = 0.05$ (corresponding to a two-tailed test), the critical region would consist of the extreme 5% of the distribution (2.5% in each tail). This means that if the calculated z-score falls in the top 2.5% or bottom 2.5% of the distribution, the null hypothesis would be rejected. In a t-test for a population mean with a sample size of 20 and a significance level of $\alpha = 0.05$, the critical values would be determined from the t-distribution table for degrees of freedom ($df = n-1 = 19$). The critical values would correspond to the 2.5% and 97.5% percentiles of the t-distribution. The critical region is important because it provides a clear criterion for deciding whether to reject the null hypothesis based on the calculated test statistic. If the test statistic falls within the critical region, the null hypothesis is rejected in favour of the alternative hypothesis. If it falls outside the critical region, the null hypothesis is not rejected.

Step 6: Compute the Test Statistic

Calculate the value of the test statistic using the sample data and the formula for the chosen test statistic. Computing the test statistic is a vital step in hypothesis testing as it quantifies the difference between the observed data and what would be expected under the null hypothesis. The choice of test statistic depends on the hypothesis being tested, the type of data, and the sample size. Here are examples illustrating how to compute the test statistic for different scenarios:

Example for a One-Sample z-test

Null Hypothesis (H_0): The population mean is 50.

Alternative Hypothesis (H_a): The population mean is not 50.

Sample Mean ($\bar{x}$): 52, Sample Standard Deviation (s): 5, Sample Size (n): 100.

$$\text{Formula}: z \frac{\bar{x}-\mu}{\frac{s}{\sqrt{n}}}$$

$$\text{Calculation}: z = \frac{52-50}{\frac{5}{\sqrt{100}}} = \frac{2}{0.5} = 4$$

Interpretation: If the calculated z-value exceeds the critical value (e.g., ±1.96 for a two-tailed test with $\alpha = 0.05$), the null hypothesis would be rejected. Here the null hypothesis is rejected. So, we can accept the alternative hypothesis, that the population mean and the sample mean are different.

Example for a Paired Sample t-test

Null Hypothesis (H_0): The mean difference between two related groups is zero.

Alternative Hypothesis (H_a): The mean difference is not zero.

Sample Mean Difference ($\bar{d}$): 1.5

Standard Deviation of Differences (s_d): 0.8

Sample Size (n): 30.

$$\text{Formula}: t = \frac{\bar{d}}{\frac{sd}{\sqrt{n}}}$$

$$\text{Calculation}: t = \frac{1.5}{\frac{0.8}{\sqrt{30}}} = \frac{1.5}{0.146} = 10.27$$

Interpretation: If the calculated t-value exceeds the critical value from the t-distribution, the null hypothesis would be rejected. Therefore, we can conclude that the null hypothesis is not supported, and an alternative hypothesis that there is a difference between the two sample means is supported.

Step 7: Make a Decision

Compare the calculated test statistic to the critical values from the distribution. If the test statistic falls in the critical region, reject the null hypothesis; otherwise, fail to reject the null hypothesis. Making a decision involves comparing the calculated test statistic to critical values from the distribution. The critical values are determined based on the chosen significance level (α) and the distribution of the test statistic. If the test statistic falls within the critical region, defined by the critical values, the null hypothesis is rejected, indicating that the observed data are unlikely to have occurred under the assumption of the null hypothesis. On the other hand, if the test statistic falls outside the critical region, the null hypothesis is not rejected, suggesting that the observed data are consistent with the null hypothesis. This decision-making process ensures that hypothesis testing is conducted systematically and allows researchers to draw valid conclusions based on the observed data and the specified significance level.

Step 8: Draw a Conclusion

Based on the decision in step 7, draw a conclusion about the null hypothesis. If you reject the null hypothesis, you accept the alternative hypothesis. If you fail to reject the null hypothesis, you do not have enough evidence to support the alternative hypothesis. If the null hypothesis is rejected, it means that there is enough evidence to support the alternative hypothesis. Conversely, if the null hypothesis is not rejected, it indicates that there is not enough evidence to support the alternative hypothesis.

For example, consider a study where a researcher wants to test if a new drug reduces blood pressure compared to a placebo. The null hypothesis (H_0) is that there is no difference in blood pressure between the two groups, while the alternative hypothesis (H_a) is that the new drug reduces blood pressure. After conducting the hypothesis test, if the p-value is less than the chosen significance level (α), the researcher would reject the null hypothesis. In this case, the conclusion would be that there is enough evidence to support the alternative hypothesis, indicating that the new drug does reduce blood pressure compared to the placebo. On the other hand, if the p-value is greater than α, the researcher would fail to reject the null hypothesis. The conclusion would then be that there is not enough evidence to support the alternative hypothesis, suggesting that the new drug may not have a significant effect on reducing blood pressure. Drawing such a conclusion after hypothesis testing is essential as it helps researchers communicate the findings of their study and make informed decisions based on the results of the hypothesis test.

Interpret the Results: Finally, interpret the results in the context of the study and make any necessary conclusions or recommendations based on the hypothesis test. Interpreting results means understanding what the statistical findings imply for the study. It goes beyond accepting or rejecting the null hypothesis, requiring a deeper analysis of the results' meaning in the context of the research question. For example, in a study comparing teaching methods, a significant difference in student performance suggests that one method is more effective. On the other hand, if there's no significant difference, it implies both methods are equally effective. The interpretation should consider practical implications and guide decisions or recommendations based on the findings.

Table 10.5 Summary Table of Steps of Hypotheses Testing

Step	Major Components	Potential Hazards
1. Formulate H_0	Define the null hypothesis (default assumption) State that there is no effect, difference, or relationship	Failing to clearly define the null hypothesis Not specifying the population parameter being tested\
2. Formulate Ha	Define the alternative hypothesis (opposite of H_0) State what you are trying to find evidence for	Formulating a vague or unclear alternative hypothesis Failing to consider all possible alternative hypotheses
3. Choose α	Select the significance level (e.g., 0.05) Determine the probability of rejecting H_0 when it's true	Choosing an inappropriate significance level Not adjusting α for multiple comparisons
4. Select Test Statistic	Choose the appropriate test statistic Determine based on hypothesis, data type, and sample size	Selecting an incorrect test statistic Using a test statistic that is not suitable for the data
5. Determine Critical Region	Identify critical values or rejection region Determine based on α and test statistic distribution	Incorrectly determining the critical values Failing to account for one-tailed vs. two-tailed tests
6. Compute Test Statistic	Calculate the test statistic using sample data	Errors in data calculation

7. Make a Decision	Compare the test statistic to critical values Decide to reject or fail to reject H_0 based on α	Misinterpreting the results Failing to consider the context of the study
8. Draw a Conclusion	Accept or reject H_0 based on the decision Interpret the findings in the context of the study	Drawing unwarranted conclusions from the results. Overgeneralizing the results to other populations

Flow Diagram for Hypothesis Testing

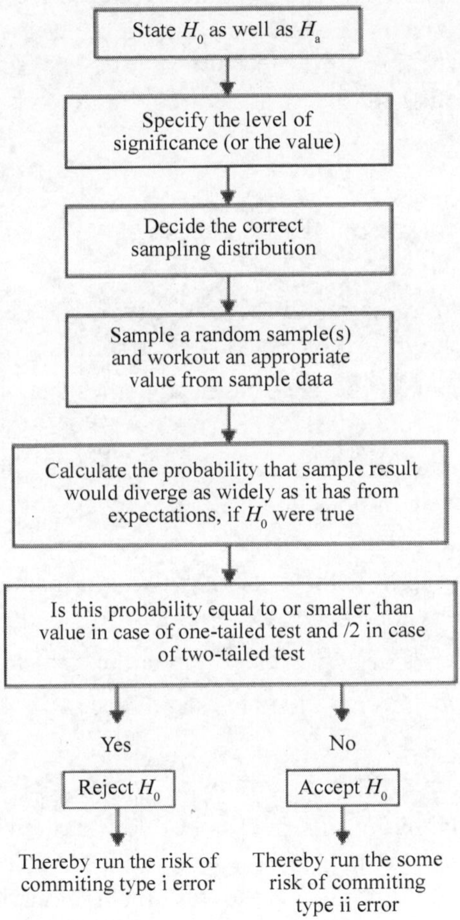

Source: Kothari, C. R. (2004). Research methodology: Methods and techniques. New Age International, p 194

10.11 Summary

In this chapter, the fundamental concepts of hypothesis testing were explored, including the definition and types of hypotheses, such as null and alternative hypotheses. The distinction between one-tailed and two-tailed tests was discussed, along with their applications in research. The chapter also examined the implications of Type I and Type II errors, emphasizing the need to understand and minimize these errors. The significance level (α) and its connection to confidence limits were discussed, showing how α influences decision-making in hypothesis testing. Additionally, the concept of degrees of freedom in hypothesis testing was explained, highlighting its role in determining the reliability of statistical results. Finally, a stepwise approach to hypothesis testing was introduced, offering a structured framework for effective hypothesis testing.

10.12 Self-Assessment

1. Define a null hypothesis and an alternative hypothesis. How are they related in hypothesis testing?
2. What is the difference between a one-tailed test and a two-tailed test? Provide an example where each type of test would be appropriate.
3. Explain the concept of Type I and Type II errors in hypothesis testing. How do they impact the validity of the test?
4. How does the significance level (α) relate to the confidence limits in hypothesis testing? Why is it important to choose an appropriate significance level?
5. Describe the stepwise approach to hypothesis testing. What are the key steps involved in conducting a hypothesis test?
6. Understand the concepts and types of hypotheses in hypothesis testing.
7. Differentiate between one-tailed and two-tailed tests and recognize their applications.
8. Understand the concept of degrees of freedom in hypothesis testing.
9. A pharmaceutical company claims that its new pain relief medication provides, on average, 30 minutes of pain relief. A study is conducted where 50 participants are randomly selected to test the medication, and their pain relief duration is measured. The sample mean is found to be 25 minutes, with a standard deviation of 10 minutes. Based on this information, what can be concluded about the company's claim regarding the average pain relief duration provided by their medication?

10. A fast-food chain claims that their drive-thru service takes, on average, 3 minutes per customer. To test this claim, a random sample of 40 drive-thru transactions is selected, and the time taken for each transaction is recorded. The sample mean time is found to be 3.5 minutes, with a standard deviation of 0.8 minutes. Based on this data, can the fast-food chain's claim be supported?
11. An online retailer claims that their website has an average uptime of 99.9% per month. To verify this claim, the website's uptime is monitored over a month, and it is found that there were 3 hours of downtime during this period. With 720 hours in a month, does this data support the retailer's claim?
12. Write null and alternative hypotheses for the following problems:
    a) A company introduces a new training program for employees to improve customer service skills. The company wants to test if the new training program leads to a significant improvement in customer satisfaction ratings. What are the null and alternative hypotheses for testing the effectiveness of the new training program on customer satisfaction ratings?
    b) A manufacturer of energy drinks claims that their product increases energy levels in consumers. To test this claim, a study is conducted to compare energy levels before and after consuming the energy drink. What are the null and alternative hypotheses for testing the effectiveness of the energy drink in increasing energy levels?
    c) A city council implements a new recycling program and wants to determine if the program has led to a significant reduction in household waste. The council collects data on household waste production before and after the program's implementation. What are the null and alternative hypotheses for testing the effectiveness of the recycling program in reducing household waste?

Unit-XI

Probability Distribution

11.1 Learning Objectives

After comprehending this unit, we will be capable to:

i) Understand the conception of probability distribution;

ii) Appreciate the significance and applications of Binomial distribution;

iii) Understand the importance and uses of Poisson distribution;

iv) Know about the normal distribution and its' applications.

11.2 Introduction

A probability distribution is a mathematical function that describes the likelihood of obtaining the possible outcomes of a random variable. In other words, it provides the probabilities of different events or values that a random variable can take. The sum of all probabilities in a probability distribution is equal to 1.

There are two main types of probability distributions: discrete and continuous.

Discrete Probability Distribution

It deals with random variables that can take on a finite or countable infinite number of distinct values. The probability mass function (PMF) is used to describe the probabilities of each possible outcome.

Example: The probability distribution of the outcomes when rolling a fair six-sided die.

Continuous Probability Distribution

It deals with random variables that can take on any value within a given range. The probability density function (PDF) is used to describe the probabilities associated with different intervals of values.

Example: The normal distribution, which is often used to model continuous random variables in various fields.

Some common probability distributions include the Binomial distribution, Poisson distribution, Exponential distribution, Uniform distribution, and many others. The choice of distribution depends on the nature of the random variable and the characteristics of the problem being analyzed.

11.3 Meaning of Probability Distribution

The term "probability distribution" refers to a statistical function or a mathematical description that represents the likelihood of various outcomes in a random experiment. It provides a systematic way of assigning probabilities to different events or values that a random variable contains. The probability distribution encompasses all possible outcomes and their associated probabilities. The probability distribution of a random variable describes how the probabilities of the outcomes of an experiment are distributed over the values of a random variable.

For a discrete random variable, the probability distribution is often described using a probability mass function (PMF), which assigns probabilities to each possible value. For a continuous random variable, the probability distribution is described using a probability density function (PDF), which gives the probability of the variable falling within a particular range.

Understanding the probability distribution is crucial for making predictions, analyzing data, and making informed decisions under uncertainty. It provides a formal and mathematical way to characterize the uncertainty associated with random variables in various scenarios.

Example 1: If two dice are rolled, then what will be the probability distribution of the sum of the dice?

Solution: The possibilities of the sum that can be obtained are (2, 3, 4, 5, 6, 7, 8, 9, 10, 11, and 12).

Suppose 1 comes up on the first die and 1 comes up on the second die. Then the sum will be 2. As no other combination of numbers give the same sum, thus, the probability of getting the sum as 2 = 1 / 36. A similar approach is followed for other numbers.

The probability distribution is given as follows:

x	2	3	4	5	6	7	8	9	10	11	12
P(x)	1 /36	2 /36	3 /36	4 /36	5 /36	6 /36	5 /36	4 /36	3 /36	2 /36	1 /36

Example 2: Suppose a die is tossed multiple times. What is the probability that the die will land on a number smaller than 6?

Solution: $P(X < 6) = P(X = 1) + P(X = 2) + P(X = 3) + P(X = 4) + P(X = 5)$

$P(X < 6) = 1 / 6 + 1 / 6 + 1 / 6 + 1 / 6 + 1 / 6 = 5 / 6$

11.4 Binomial Distribution

The binomial distribution is a discrete probability distribution that describes the number of successes in a fixed number of independent Bernoulli trials, where each trial has only two possible outcomes: success or failure. The outcomes are often denoted as "success" and "failure," but they can represent any dichotomous events, such as yes/no, true/false, or heads/tails.

The key characteristics of a binomial distribution are:

i) **Fixed Number of Trials (n):** The number of trials is predetermined and remains constant throughout the experiment.

ii) **Independent Trials:** The outcome of one trial does not affect the outcome of any other trial. Each trial is considered independent.

iii) **Constant Probability of Success (p):** The probability of success (denoted as p) remains constant for each trial.

iv) **Discrete Outcomes:** The random variable representing the number of successes takes on discrete values (0, 1, 2, ..., n).

The probability mass function (PMF) of the Binomial distribution can be expressed as:

$$P(X=k)=\binom{n}{k}.p^k.(1-p)^{n-k}$$

Where, is the probability of getting exactly k successes in n trials; $\binom{n}{k}$ is the binomial coefficient, representing the number of ways to choose k successes from n trials; is the probability of success on a single trial; (1 - p) is the probability of failure on a single trial; is the total number of trials; and is the number of successes.

The mean (expected value) and standard deviation of a binomial distribution are given by:

Mean (μ) = np

Standard Deviation $(\sigma)=\sqrt{np(1-p)}$

The Binomial distribution is generally used in various fields, such as quality control, genetics, financial analysis, and statistics, to model the number of successes in a fixed number of trials with binary outcomes.

Examples

Question 1

Find the binomial distribution of getting a six in three tosses of an unbiased dice.

Solution

Let X be the random variable of getting six. Then X can be 0, 1, 2, 3.

Here, n = 3

p = Probability of getting a six in a toss = 1/6

q = Probability of not getting a six in a toss = 1 – 1/6 = 5/6

$P(X = 0) = {}^nC_r\, p^r\, q^{(n-r)} = {}^3C_0\, (1/6)^0\, (5/6)^{3-0}$

$= 1 \times 1 \times 125/216 = 125/216$

$P(X = 1) = {}^nC_r\, p^r\, q^{(n-r)} = {}^3C_1\, (1/6)^1\, (5/6)^{3-1}$

$= 3 \times 1/6 \times 25/36 = 25/72$

$P(X = 2) = {}^nC_r\, p^r\, q^{(n-r)} = {}^3C_2\, (1/6)^2\, (5/6)^{3-2}$

$= 3 \times 1/36 \times 5/6 = 5/72$

$P(X = 3) = {}^nC_r\, p^r\, q^{(n-r)} = {}^3C_3\, (1/6)^3\, (5/6)^{3-3}$

$= 1 \times 1/216 \times 1 = 1/216$

The required binomial distribution of X is:

X	**0**	**1**	**2**	**3**
p(X)	**125/216**	**25/72**	**5/72**	**1/216**

Question 2

Find the probability distribution of the number of doublets in four throws of a pair of dice.

Solution

There are 36 total possible outcomes for a throw of dice, for which the following outcomes are the success of the experiment: {(1,1), (2, 2), (3, 3), (4, 4), (5, 5), (6, 6)}

p = probability of getting doublets = 6/36 = 1/6

q = probability of getting not getting doublets = 1 – 1/6 = 5/6

X: numbers of doublets, then X = 0, 1, 2, 3, and 4.

$P(X = 0) = {}^nC_r\, p^r\, q^{(n-r)} = {}^4C_0\, (1/6)^0\, (5/6)^{4-0}$

$= 1 \times 1 \times 625/1296$

$P(X = 1) = {}^nC_r\, p^r\, q^{(n-r)} = {}^4C_1\, (1/6)^1\, (5/6)^{4-1}$

$= 4 \times 1/6 \times 125/216 = 125/324$

$P(X = 2) = {}^nC_r\, p^r\, q^{(n-r)} = {}^4C_2\, (1/6)^2\, (5/6)^{4-2}$

$= 6 \times 1/36 \times 25/36$

$= 25/216$

$P(X = 3) = {}^nC_r\, p^r\, q^{(n-r)} = {}^4C_3\, (1/6)^3\, (5/6)^{4-3}$

$= 4 \times 1/216 \times 5/6 = 20/1296$

$P(X = 4) = {}^nC_r\, p^r\, q^{(n-r)} = {}^4C_4\, (1/6)^4\, (5/6)^{4-4}$

$= 1 \times 1/1296 \times 1 = 1/1296.$

\ The required probability distribution is:

X	0	1	2	3	4
P(X)	625/1296	125/324	25/216	20/1296	1/1296

Question 3

Find the probability of getting at least 5 times head-on tossing an unbiased coin for 6 times by using the binomial distribution.

Solution

p = P (getting an head in a single toss) = ½

q = P (not getting an head in a single toss) = ½

X = successfully getting a head

$P(X \geq 5)$ = P (getting at least 5 heads) = $P(X = 5) + P(X = 6)$

= 6C5 (½)5 (½) (6 – 5) + 6C6 (½)6 (½)6 – 6

= 6 × (½)6 + 1 × (½)6 = 7/24.

Hence, the probability of getting at least 5 heads is 7/24.

Question 4

On average, every one out of 10 telephones is found busy. Six telephone numbers are selected at random. Find the probability that four of them will be busy.

Solution

Let X: event of getting a busy phone number

p = P (probability of getting a phone number busy) = 1/10

q = P (probability of not getting a phone number busy) = 9/10

The required probability = $P(X = 4) = {}^6C_4\, p^4\, q^{(6-4)}$

$= 15 \times (1/10)^4 \times (9/10)^2$

$= 15 \times 81/10^6$

$= 0.001215.$

Question 5

An unbiased dice is thrown until three sixes are obtained. Find the probability of obtaining the third six in the sixth throw.

Solution

Since each throw is independent of the previous throws, we can apply the binomial distribution formula to find the probability.

$p = P$ (getting a six in a throw) $= 1/6$

$q = P$ (not getting a six in a throw) $= 1 - 1/6 = 5/6$

According to the question, two sixes are already obtained in the previous throws.

\Required probability = P (getting exactly two sixes in five throws) × P (getting a six in the sixth throw)

$= {}^5C_2\, p^2q^3 \times {}^1C_1 p^1\, q^{1-1}$

$= 10 \times (1/6)^2 \times (5/6)^3 \times 1 \times (1/6)$

$= 10 \times (1/6)^3 \times (5/6)^3$

$= 625/23328.$

Question 6

When a biased coin is tossed, the probability of getting a head 3 times more than the probability of getting a tail. Find the probability distribution for getting a tail, if the coin is tossed twice.

Solution

Let the probability of getting a tail be p, then the probability of getting a head will be 3p

Now, $p + 3p = 1 \rightarrow p = \frac{1}{4}$

$q = P$ (not getting a tail) $= 1 - \frac{1}{4} = \frac{3}{4}$

X = event of getting a tail in a toss

Then, possible values of x will be 0, 1, and 2

$P(X = 0) = {}^2C_0\, p^0\, q^{2-0}$

$= 1 \times 1 \times (\frac{3}{4})^2$

$= 9/16$

$P(X = 1) = {}^2C_1\, p^1\, q^{2-1}$

$= 2 \times (\frac{1}{4}) \times (\frac{3}{4})$

$= \frac{3}{8}$

$P(X = 2) = {}^2C_2\, p^2\, q^{2-2}$

$= 1 \times (\frac{1}{4})^2 \times 1$

$= 1/16$

The probability distribution for getting the tail is:

X	0	1	2
P(X)	9/16	3/8	1/16

Question 7

A bag contains 5 green balls and 3 red balls. If two balls are drawn from the bag randomly with replacement, find the probability distribution of the number of green balls drawn.

Solution

Let p = P (getting a green ball) = 5 (5 + 3) = 5/8

q = P = (not getting a green ball) = 1 – 5/8 = 3/8

X = event of drawing the green ball, then the value of X could be 0, 1, 2

P(X = 0) = Probability of getting no green ball = $^2C_0\, p^0\, q^{2-0} = 1 \times 1 \times (3/8)^2 =$ 9/64

P(X = 1) = Probability of getting one green ball = $^2C_1\, p^1\, q^{2-1} = 2 \times (5/8) \times (3/8)$ = 15/32

P(X = 2) = Probability of getting 2 green balls = $^2C_2\, p^2\, q^{2-2} = 1 \times (5/8)^2 \times (3/8)^0 =$ 25/64

The required probability distribution is:

X	0	1	2
P(X)	9/64	15/32	25/64

Question 8

Find the probability distribution of getting the number of fours in three throws of a dice. Also, find the mean and variance of the distribution.

Solution

Let, p = P (getting a four in a throw of dice) = 1/6

q = P (not getting a four in a throw of dice) = 5/6

X: number of four obtained, then the value of X could be 0, 1, 2, and 3.

$P(X = 0) = {}^3C_0\, p^0q^{3-0} = 1 \times (5/6)^3 = 125/216$

$P(X = 1) = {}^3C_1\, p^1q^{3-1} = 3 \times (1/6) \times (5/6)^2 = 75/216$

$P(X = 2) = {}^3C_2\, p^2q^{3-2} = 1 \times (1/6)^2 \times (5/6)^{3-2} = 15/216$

$P(X = 3) = {}^3C_3\, p^3q^{3-3} = 1 \times (1/6)^3 \times (5/6)^{3-3} = 1/216$

The required probability distribution

X	0	1	2	3
P(X)	125/216	75/216	15/216	1/216

Mean = np = 3 × 1/6 = 1.2

Variance = npq = 3 × 1/6 × 5/6 = 5/12.

11.5 Poisson Distribution

Poisson distribution is actually another probability distribution formula. As per binomial distribution, we won't be given the number of trials or the probability of success on a certain trial. The average number of successes will be given in a certain time interval. The average number of successes is called "Lambda" and denoted by the symbol "λ".

The formula for Poisson distribution formula is given below:

$$P(X=x)=\frac{e^{-\lambda}\lambda^{x}}{x!}$$

Here, λ is the average number,

x is a Poisson random variable,

e is the base of logarithm and $e = 2.71828$ (approx.).

Mean and variance of Poisson Distribution

If μ is the average number of successes occurring in a given time interval or region in the Poisson distribution, then the mean and the variance of the Poisson distribution are both equal to μ.

$E(X) = \mu$

and $V(X) = \sigma^2 = \mu$

Note: In a Poisson distribution, only one parameter, μ is needed to determine the probability of an event.

Examples

Question 1: As only 3 students came to attend the class today, find the probability for exactly 4 students to attend the classes tomorrow.

Solution

Given,

Average rate of value (μ) = 3

Poisson random variable(x) = 4

Poisson distribution =

$$P(X=x)=\frac{e^{-\lambda}\lambda^{x}}{x!}$$

$$P(X=4)=\frac{e^{-3},3^{4}}{4!}$$

$P(X = 4) = 0.16803135574154$

Question 2: If the random variable X follows a Poisson distribution with mean 3.4, find P(X=6).

Solution

This can be written more quickly as: if X ~ Po(3.4) find P(X=6).

The Poisson distribution =

$$P(X=6)=\frac{e^{-\lambda}\lambda^{6}}{6!}$$

$$=\frac{e^{-3.4}(3.4)^{6}}{6!}(\text{mean}, \lambda=3.4)$$

= 0.071604409 = 0.072 (to 3 d. p.)

Question 3: The number of industrial injuries per working week in a particular factory is known to follow a Poisson distribution with mean 0.5. Find the probability that

a) In a particular week there will be:

i) Less than 2 accidents,

ii) More than 2 accidents;

b) In a three-week period there will be no accidents.

Solution

Let A be 'the number of accidents in one week', so $A \sim P_0(0.5)$.

(a) (i) P (A< 2) = P (A = 1)

= 0.9098

Or from the formula,

P(A < 2) = P (A = 0) + P (A = 1)

$$= e^{-0.5} + \frac{e^{-0.5} \times 0.5}{1!}$$

$$= \frac{3}{2}e^{-0.5}$$

~ 0.9098

(b) (ii) $P(A > 2) = 1 - P(A < 2)$

$= 1 - 0.9856$ (*from tables*)

$= 0.0144$ (to 4 *d.p.*)

Or

$1 - [P(A = 0) + (A = 1) + P(A = 2)$

$$=1-\left[e^{-0.5}+e^{-0.5}+\frac{e^{-0.5}(0.5)^2}{2!}\right]$$

$= 1 - e\text{-}0.5\ (1 + 0.5 + 0.125)$

≈ 0.0144

$P(0 \text{ in } 3 \text{ weeks}) = (e^{-0.5})3 \approx 0.223$

Question 4: In a cafe, the customer arrives at a mean rate of 2 per min. Find the probability of arrival of 5 customers in 1 minute using the Poisson distribution formula.

Solution

Given: $\lambda = 2$, and $x = 5$.

Using the Poisson distribution formula:

$P(X = x) = (e^{-\lambda} \lambda^x)/x!$

$P(X = 5) = (e^{-2} 2^5)/5!$

$P(X = 6) = 0.036$

The probability of arrival of 5 customers per minute is 3.6%.

Question 5: Find the mass probability of function at $x = 6$, if the value of the mean is 3.4.

Solution

Given: $\lambda = 3.4$, and $x = 6$.

Using the Poisson distribution formula:

$P(X = x) = (e^{-\lambda} \lambda^x)/x!$

$P(X = 6) = (e^{-3.4} 3.4^6)/6!$

$P(X = 6) = 0.072$

The probability of function is 7.2%.

Question 6: If 3% of electronic units manufactured by a company are defective. Find the probability that in a sample of 200 units, less than 2 bulbs are defective.

Solution

The probability of defective units p = 3/100 = 0.03

Give n = 200.

We observe that p is small and n is large here. Thus, it is a Poisson distribution.

Mean $\lambda = np = 200 \times 0.03 = 6$

P(X= x) is given by the Poisson distribution formula as $(e^{-\lambda}\lambda^{x})/x!$

$P(X < 2) = P(X = 0) + P(X = 1)$

$= (e^{-6}\,6^{0})/0! + (e^{--6}6^{1})/1!$

$= e^{-6} + e^{--6} \times 6$

$= 0.00247 + 0.0148$

$P(X < 2) = 0.01727$

The probability that less than 2 bulbs are defective is 0.01727.

11.6 Normal Distribution

The normal (or Gaussian) distribution is the most common continuous probability distribution. The function gives the probability than an event will fall between any two real number limits as the curve approaches zero on either side of the mean. Area underneath the normal curve is always equal to 1.

The curve itself is approximately bell shaped, and is therefore informally called the Bell Curve.

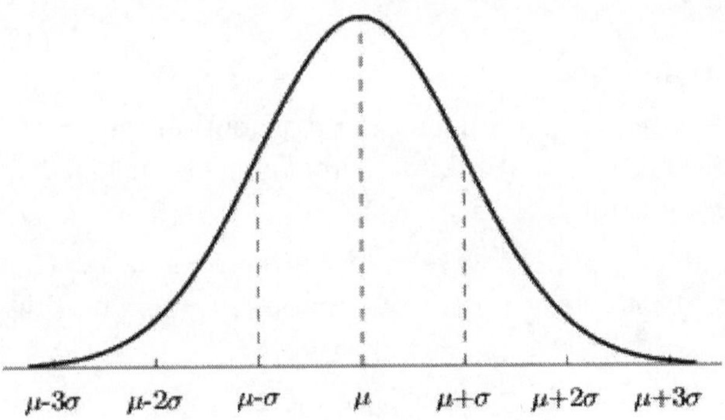

$$f(x)=\frac{1}{\sigma\sqrt{2\pi}}e^{\frac{-(x-\mu)^2}{2\sigma^2}}$$

For the normal curve, the mean, median, and mode should all be the same value (μ). The normal curve also has the axis of symmetry . For the standard normal curve, and .

Empirical Rule

Generally, for a Normal Curve, 68% of all values of x fall between $\mu \pm 1\sigma$; 95% of all values of x will fall between $\mu \pm 2\sigma$; and 99.7% of all values of x will fall between $\mu \pm 3\sigma$.

The Central Limit Theorem

One of the aspects of the normal curve that makes it so powerful is the central limit theorem, which states that for a sufficiently large number of samples, the mean of these many samples will be normally distributed, regardless of the distribution of the original data or events.

The **standard normal distribution** is one of the forms of the normal distribution. It occurs when a normal random variable has a mean equal to zero and a standard deviation equal to one. In other words, a normal distribution with a mean 0 and standard deviation of 1 is called the standard normal distribution. Also, the standard normal distribution is centred at zero, and the standard deviation gives the degree to which a given measurement deviates from the mean.

The random variable of a standard normal distribution is known as the **standard score or a z-score**. It is possible to transform every normal random variable X into a z score using the following formula:

$$\mathbf{z = (X - \mu) / \sigma}$$

Where X is a normal random variable, μ is the mean of X, and σ is the standard deviation of X.

Properties of Normal Distribution

The normal distribution, also known as the Gaussian distribution, possesses several important properties, which make it a fundamental concept in statistics and probability theory. Here are some key properties:

i) **Symmetry:** The normal distribution is symmetric around its mean. This means that the probability of observing a value to the left or right of the mean is the same.

ii) **Bell-shaped Curve:** The probability density function of the normal distribution forms a bell-shaped curve, with the highest point at the mean.

iii) **Mean, Median, and Mode Equality:** The mean (average), median, and mode of a normal distribution are all equal and located at the center of the distribution.

iv) **Parameterized by Mean and Standard Deviation:** The normal distribution is completely characterized by two parameters: the mean (μ) and the standard deviation (σ). The mean determines the location of the center, and the standard deviation controls the spread or width of the distribution.

v) **68-95-99.7 Rule:** This rule states that approximately 68% of the data falls within one standard deviation of the mean, about 95% within two standard deviations, and approximately 99.7% within three standard deviations.

vi) **Empirical Rule:** Similar to the 68-95-99.7 rule, the empirical rule states that in a normal distribution, about 68% of the data falls within one standard deviation, 95% within two standard deviations, and 99.7% within three standard deviations.

vii) **Central Limit Theorem:** One of the most important properties of the normal distribution is its connection to the Central Limit Theorem. According to this theorem, the sum or average of a large number of independent, identically distributed random variables approaches a normal distribution, regardless of the original distribution of the variables.

viii) **Additivity of Normal Random Variables:** If X and Y are independent normal random variables, then their sum $(X+Y)$ is also a normal random variable.

ix) **Z-Score:** A Z-score, also known as a standard score, measures how many standard deviations a data point is from the mean of a normal distribution. It is calculated as $(X - \mu) / \sigma$, where X is the data point, μ is the mean, and σ is the standard deviation.

Examples

Problem 1: For some computers, the time period between charges of the battery is normally distributed with a mean of 50 hours and a standard deviation of 15 hours. Rohan has one of these computers and needs to know the probability that the time period will be between 50 and 70 hours.

Solution

Let x be the random variable that represents the time period.

Given Mean, $\mu = 50$

and standard deviation, $\sigma = 15$

To find: Probability that x is between 50 and 70 or $P(50 < x < 70)$

By using the transformation equation, we know;

$z = (X - \mu) / \sigma$

For x = 50 , z = (50 – 50) / 15 = 0

For x = 70 , z = (70 – 50) / 15 = 1.33

P(50< x < 70) = P(0< z < 1.33) = [area to the left of z = 1.33] – [area to the left of z = 0]

From the table we get the value, such as;

P(0 < z < 1.33) = 0.9082 – 0.5 = 0.4082

The probability that Rohan's computer has a time period between 50 and 70 hours is equal to 0.4082.

Problem 2: The speeds of cars are measured using a radar unit, on a motorway. The speeds are normally distributed with a mean of 90 km/hr and a standard deviation of 10 km/hr. What is the probability that a car selected at chance is moving at more than 100 km/hr?

Solution

Let the speed of cars is represented by a random variable 'x'.

Now, given mean, $\mu = 90$ and standard deviation, $\sigma = 10$.

To find: Probability that x is higher than 100 or P(x > 100)

By using the transformation equation, we know;

$z = (X - \mu) / \sigma$

Hence,

For x = 100 , z = (100 – 90) / 10 = 1

P(x > 90) = P(z > 1) = [total area] – [area to the left of z = 1]

P(z > 1) = 1 – 0.8413 = 0.1587

Problem 3: Most graduate schools of business require applicants for admission to take the Graduate Management Admission Council's GMAT examination. Scores on the GMAT are roughly normally distributed with a mean of 527 and a standard deviation of 112. What is the probability of an individual scoring above 500 on the GMAT?

Solution

$$Z = \frac{500-527}{112} = -0.24107$$

Normal Distribution

$\mu = 527, \alpha = 112,$

Pr {X > 500} = Pr {Z > - 0.24} = 1 - 0.4052 = 0.5948

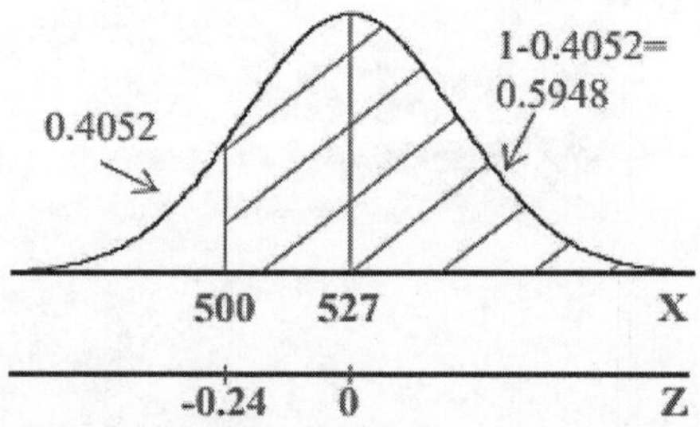

Problem 4: How high must an individual score on the GMAT in order to score in the highest 5%?

Solution

Normal Distribution

μ=527, α=112

$P(X > ?) = 0.05 => P(Z > ?) = 0.05$

$P(Z > ?) => 1- 0.05 = 0.95 => Z=1.645$

$X= 527 + 1.645(112)$

$X= 527 + 184.24$

$X= 711.24$

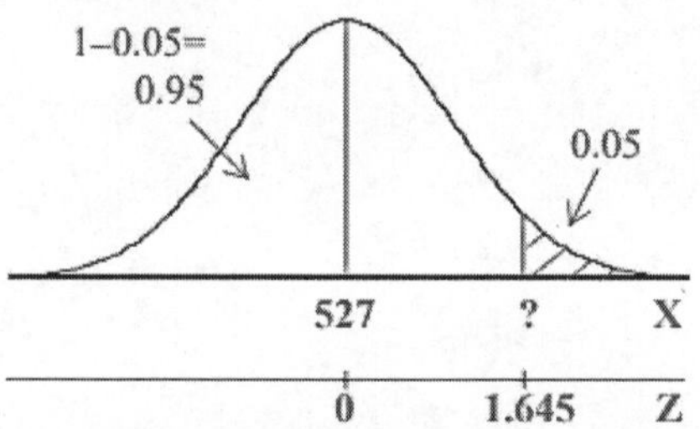

Problem 5: The length of human pregnancies from conception to birth approximates a normal distribution with a mean of 266 days and a standard deviation of 16 days. What proportion of all pregnancies will last between 240 and 270 days (roughly between 8 and 9 months)?

Solution

$$Z = \frac{240-266}{16} = -1.625$$

$$Z = \frac{270-266}{16} = 0.25$$

Normal Distribution

μ=266,

α=16

P (240 < X < 270) = P (-1.63 < Z < 0.25)

P (-1.63 < Z < 0.25) = P (Z < 0.25) - P (Z < -1.63)

P (-1.63 < Z < 0.25) = 0.5987-0.0516 = 0.5471

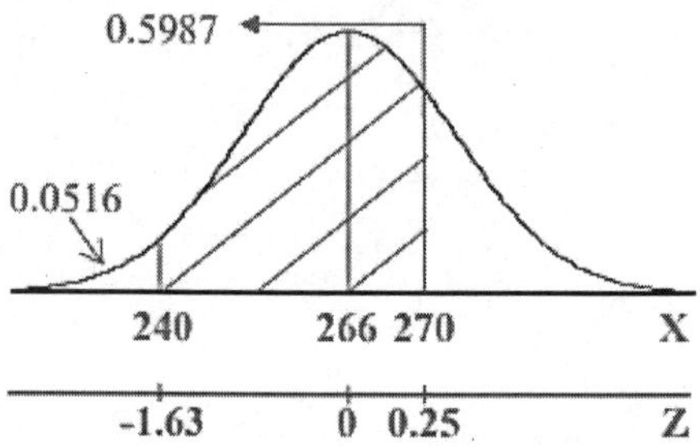

Problem 6: The average number of acres burned by forest and range fires in a large New Mexico country is 4,300 acres per year, with a standard deviation of 750 acres. The distribution of the number of acres burned is normal. What is the possibility that between 2,500 and 4200 acres will be burned in any given year?

Solution

$$Z = \frac{240-266}{16} = -1.625$$

$$Z = \frac{270-266}{16} = 0.25$$

Normal Distribution

μ = 4300

α = 750

P(2500<X<4200)=P(-2.40<Z<-0.13)

P(-2.40<Z<-0.13)=P(<-0.13)-P(Z<-2.40)

P(-2.40<Z<-0.13)=0.4483-0.0082=0.4401

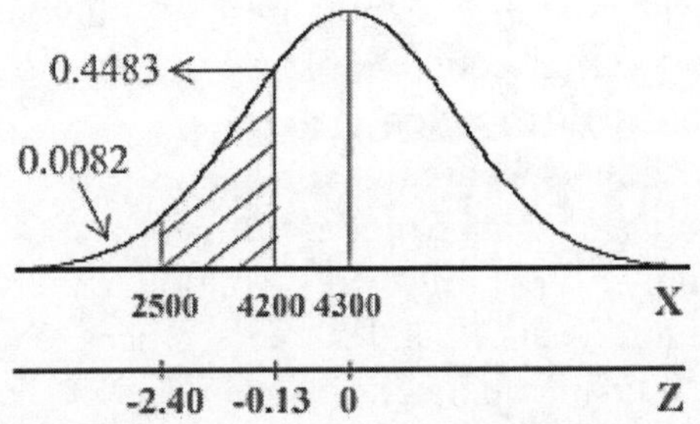

Problem 7: The Edwards's Theater chain has studied its movie customers to determine how much money they spend on concessions. The study revealed that the spending distribution is approximately normally distributed with a mean of \$4.11 and a standard deviation of \$1.37. What percentage of customers will spend less than \$3.00 on concessions?

Solution

$$Z = \frac{3.00 - 4.11}{1.37} = -0.81021$$

Normal Distribution

$\mu = 4.11$

$\alpha = 1.37$

P (X < 3.00) = P (Z < -0.81) = 0.2090

= 20.9%

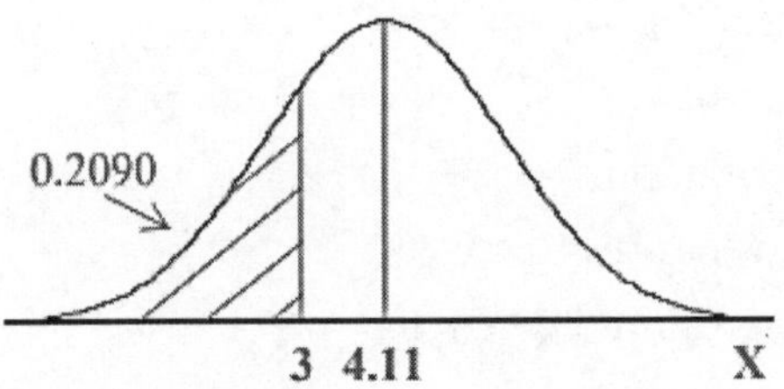

11.7 Summary

The chapter offers a thorough comprehension of probability distribution and its various types. The chapter presents the fundamental concepts of probability distribution. A probability distribution is a mathematical function that provides information about the probability of various outcomes of a random variable. It is an essential instrument for examining ambiguity. The chapter's progression continued by exploring and acknowledging the significance of the Binomial distribution. The discrete probability distribution is crucial in situations where we want to determine the number of successful outcomes in a set number of independent Bernoulli trials. The following section explored the Poisson distribution, which is a discrete probability distribution that is highly useful for representing infrequent occurrences within specific time or space intervals. Finally, the chapter ended by providing a thorough comprehension of the normal distribution, which is a continuous probability distribution widely utilized in various fields, including natural sciences and social sciences. We identified the bell-shaped curve and its fundamental significance in statistical analysis. However, the chapter presents several examples to acquaint readers with each distribution process and its mathematical calculation procedure.

11.8 Self-Assessment

Multiple Choice Questions (MCQ)

1. What is a probability distribution?
   a) A function that describes the likelihood of obtaining the possible values that a random variable can assume.
   b) A statistical method for data collection.
   c) The average value of a dataset.
   d) A graphical representation of frequency data.
2. Which of the following is a true statement about the Binomial distribution?
   a) It describes the distribution of outcomes in experiments with only one possible outcome.
   b) It is used for describing events with two possible outcomes, like success and failure.
   c) It is the same as the Normal distribution.
   d) It cannot be used in real-world scenarios.
3. The Poisson distribution is most appropriate when:
   a) The events are dependent on each other.
   b) The number of trials is very large, and the probability of success is very small.
   c) There are only two possible outcomes.
   d) The data is continuous.
4. Which of the following is a characteristic of the Normal distribution?
   a) It is skewed to the right.
   b) It is symmetric around its mean.

c) It can only model discrete data.

d) Its mean, median, and mode are all different.

Short Answer Questions

5. Describe the concept of a probability distribution.
6. Explain why the Binomial distribution is important in statistical analysis.
7. Discuss the importance of the Poisson distribution and its typical applications.
8. What are the key characteristics of the Normal distribution?

Long Answer Questions

9. Assume that the chance of an individual coal miner being killed in a mine accident during a year is 1/1400. Use the Poisson distribution to calculate the probability that in a mine employing 350 miners, there will be at least one fatal accident in a year. (use $e^{-0.25} = 0.78$)
10. The mean and standard deviation of a Poisson distribution are 6 and 2 respectively. Test the validity of this statement.
11. A typist commits the following mistakes per page in typing 100 pages. Fit a Poisson distribution and calculate the theoretical frequencies.

Mistakes per page (X)	0	1	2	3	4	5
Frequency (f)	42	33	14	6	4	1

12. Seven coins are tossed and the number of heads is noted. The experiment is repeated 128 times and the following distribution is highlighted in the below table. Fit a binomial distribution assuming the coin is unbiased.

Number of heads	0	1	2	3	4	5	6	7
Frequencies	7	6	19	35	30	23	7	1

Appendix

Value of $e^{-\lambda}$ (For Computing Poisson Probabilities)

$(0 < \lambda < I)$

λ	0	1	2	3	4	5	6	7	8	9
0.0	1.0000	0.9900	0.9802	0.9704	0.9608	0.9512	0.9418	0.9324	0.9231	0.9139
0.1	0.9048	0.8958	0.8860	0.8781	0.8694	0.8607	0.8521	0.8437	0.8353	0.8270
0.2	0.7187	0.8106	0.8025	0.7945	0.7866	0.7788	0.7711	0.7634	0.7558	0.7483
0.3	0.7408	0.7334	0.7261	0.7189	0.7118	0.7047	0.6970	0.6907	0.6839	0.6771
0.4	0.6703	0.6636	0.6570	0.6505	0.6440	0.6376	0.6313	0.6250	0.6188	0.6125
0.5	0.6065	0.6005	0.5945	0.5886	0.5827	0.5770	0.5712	0.5655	0.5599	0.5543
0.6	0.5448	0.5434	0.5379	0.5326	0.5278	0.5220	0.5160	0.5113	0.5066	0.5016
0.7	0.4966	0.4916	0.4868	0.4810	0.4771	0.4724	0.4670	0.4630	0.4584	0.4538
0.8	0.4493	0.4449	0.4404	0.4360	0.4317	0.4274	0.4232	0.4190	0.4148	0.4107
0.9	0.4066	0.4026	0.3985	0.3946	0.3906	0.3867	0.3829	0.3791	0.3753	0.3716
(λ=1,2,3,…,10)										
λ	1	2	3	4	5	6	7	8	9	10
e-λ	0.3679	0.1353	0.0498	0.0183	0.0070	0.0028	0.0009	0.0004	0.0001	0.00004

Note: To obtain values of $e^{-\lambda}$ for other values of λ, use the laws of exponents i.e. $e^{-(a+b)} = e^{-a}. e^{-b}$. For example: $e^{-2.25} = e^{-2}. e^{-0.25} = (0.1353). (0.7788) = 0.1054$.

STANDARD NORMAL DISTRIBUTION: Table Values Represent AREA to the LEFT of the Z score.

Z	.00	.01	.02	.03	.04	.05	.06	.07	.08	.09
-3.9	.00005	.00005	.00004	.00004	.00004	.00004	.00004	.00004	.00003	.00003
-3.8	.00007	.00007	.00007	.00006	.00006	.00006	.00006	.00005	.00005	.00005
-3.7	.00011	.00010	.00010	.00010	.00009	.00009	.00008	.00008	.00008	.00008
-3.6	.00016	.00015	.00015	.00014	.00014	.00013	.00013	.00012	.00012	.00011
-3.5	.00023	.00022	.00022	.00021	.00020	.00019	.00019	.00018	.00017	.00017
-3.4	.00034	.00032	.00031	.00030	.00029	.00028	.00027	.00026	.00025	.00024
-3.3	.00048	.00047	.00045	.00043	.00042	.00040	.00039	.00038	.00036	.00035
-3.2	.00069	.00066	.00064	.00062	.00060	.00058	.00056	.00054	.00052	.00050
-3.1	.00097	.00094	.00090	.00087	.00084	.00082	.00079	.00076	.00074	.00071
-3.0	.00135	.00131	.00126	.00122	.00118	.00114	.00111	.00107	.00104	.00100
-2.9	.00187	.00181	.00175	.00169	.00164	.00159	.00154	.00149	.00144	.00139
-2.8	.00256	.00248	.00240	.00233	.00226	.00219	.00212	.00205	.00199	.00193
-2.7	.00347	.00336	.00326	.00317	.00307	.00298	.00289	.00280	.00272	.00264
-2.6	.00466	.00453	.00440	.00427	.00415	.00402	.00391	.00379	.00368	.00357
-2.5	.00621	.00604	.00587	.00570	.00554	.00539	.00523	.00508	.00494	.00480
-2.4	.00820	.00798	.00776	.00755	.00734	.00714	.00695	.00676	.00657	.00639
-2.3	.01072	.01044	.01017	.00990	.00964	.00939	.00914	.00889	.00866	.00842
-2.2	.01390	.01355	.01321	.01287	.01255	.01222	.01191	.01160	.01130	.01101
-2.1	.01786	.01743	.01700	.01659	.01618	.01578	.01539	.01500	.01463	.01426
-2.0	.02275	.02222	.02169	.02118	.02068	.02018	.01970	.01923	.01876	.01831
-1.9	.02872	.02807	.02743	.02680	.02619	.02559	.02500	.02442	.02385	.02330
-1.8	.03593	.03515	.03438	.03362	.03288	.03216	.03144	.03074	.03005	.02938
-1.7	.04457	.04363	.04272	.04182	.04093	.04006	.03920	.03836	.03754	.03673

-1.6	.05480	.05370	.05262	.05155	.05050	.04947	.04846	.04746	.04648	.04551
-1.5	.06681	.06552	.06426	.06301	.06178	.06057	.05938	.05821	.05705	.05592
-1.4	.08076	.07927	.07780	.07636	.07493	.07353	.07215	.07078	.06944	.06811
-1.3	.09680	.09510	.09342	.09176	.09012	.08851	.08691	.08534	.08379	.08226
-1.2	.11507	.11314	.11123	.10935	.10749	.10565	.10383	.10204	.10027	.09853
-1.1	.13567	.13350	.13136	.12924	.12714	.12507	.12302	.12100	.11900	.11702
-1.0	.15866	.15625	.15386	.15151	.14917	.14686	.14457	.14231	.14007	.13786
-0.9	.18406	.18141	.17879	.17619	.17361	.17106	.16853	.16602	.16354	.16109
-0.8	.21186	.20897	.20611	.20327	.20045	.19766	.19489	.19215	.18943	.18673
-0.7	.24196	.23885	.23576	.23270	.22965	.22663	.22363	.22065	.21770	.21476
-0.6	.27425	.27093	.26763	.26435	.26109	.25785	.25463	.25143	.24825	.24510
-0.5	.30854	.30503	.30153	.29806	.29460	.29116	.28774	.28434	.28096	.27760
-0.4	.34458	.34090	.33724	.33360	.32997	.32636	.32276	.31918	.31561	.31207
-0.3	.38209	.37828	.37448	.37070	.36693	.36317	.35942	.35569	.35197	.34827
-0.2	.42074	.41683	.41294	.40905	.40517	.40129	.39743	.39358	.38974	.38591
-0.1	.46017	.45620	.45224	.44828	.44433	.44038	.43644	.43251	.42858	.42465
-0.0	.50000	.49601	.49202	.48803	.48405	.48006	.47608	.47210	.46812	.46414

STANDARD NORMAL DISTRIBUTION: Table Values Represent AREA to the LEFT of the Z score.

Z	.00	.01	.02	.03	.04	.05	.06	.07	.08	.09
0.0	.50000	.50399	.50798	.51197	.51595	.51994	.52392	.52790	.53188	.53586
0.1	.53983	.54380	.54776	.55172	.55567	.55962	.56356	.56749	.57142	.57535
0.2	.57926	.58317	.58706	.59095	.59483	.59871	.60257	.60642	.61026	.61409
0.3	.61791	.62172	.62552	.62930	.63307	.63683	.64058	.64431	.64803	.65173
0.4	.65542	.65910	.66276	.66640	.67003	.67364	.67724	.68082	.68439	.68793
0.5	.69146	.69497	.69847	.70194	.70540	.70884	.71226	.71566	.71904	.72240
0.6	.72575	.72907	.73237	.73565	.73891	.74215	.74537	.74857	.75175	.75490
0.7	.75804	.76115	.76424	.76730	.77035	.77337	.77637	.77935	.78230	.78524
0.8	.78814	.79103	.79389	.79673	.79955	.80234	.80511	.80785	.81057	.81327
0.9	.81594	.81859	.82121	.82381	.82639	.82894	.83147	.83398	.83646	.83891
1.0	.84134	.84375	.84614	.84849	.85083	.85314	.85543	.85769	.85993	.86214
1.1	.86433	.86650	.86864	.87076	.87286	.87493	.87698	.87900	.88100	.88298
1.2	.88493	.88686	.88877	.89065	.89251	.89435	.89617	.89796	.89973	.90147
1.3	.90320	.90490	.90658	.90824	.90988	.91149	.91309	.91466	.91621	.91774
1.4	.91924	.92073	.92220	.92364	.92507	.92647	.92785	.92922	.93056	.93189
1.5	.93319	.93448	.93574	.93699	.93822	.93943	.94062	.94179	.94295	.94408
1.6	.94520	.94630	.94738	.94845	.94950	.95053	.95154	.95254	.95352	.95449
1.7	.95543	.95637	.95728	.95818	.95907	.95994	.96080	.96164	.96246	.96327
1.8	.96407	.96485	.96562	.96638	.96712	.96784	.96856	.96926	.96995	.97062
1.9	.97128	.97193	.97257	.97320	.97381	.97441	.97500	.97558	.97615	.97670
2.0	.97725	.97778	.97831	.97882	.97932	.97982	.98030	.98077	.98124	.98169
2.1	.98214	.98257	.98300	.98341	.98382	.98422	.98461	.98500	.98537	.98574
2.2	.98610	.98645	.98679	.98713	.98745	.98778	.98809	.98840	.98870	.98899

2.3	.98928	.98956	.98983	.99010	.99036	.99061	.99086	.99111	.99134	.99158
2.4	.99180	.99202	.99224	.99245	.99266	.99286	.99305	.99324	.99343	.99361
2.5	.99379	.99396	.99413	.99430	.99446	.99461	.99477	.99492	.99506	.99520
2.6	.99534	.99547	.99560	.99573	.99585	.99598	.99609	.99621	.99632	.99643
2.7	.99653	.99664	.99674	.99683	.99693	.99702	.99711	.99720	.99728	.99736
2.8	.99744	.99752	.99760	.99767	.99774	.99781	.99788	.99795	.99801	.99807
2.9	.99813	.99819	.99825	.99831	.99836	.99841	.99846	.99851	.99856	.99861
3.0	.99865	.99869	.99874	.99878	.99882	.99886	.99889	.99893	.99896	.99900
3.1	.99903	.99906	.99910	.99913	.99916	.99918	.99921	.99924	.99926	.99929
3.2	.99931	.99934	.99936	.99938	.99940	.99942	.99944	.99946	.99948	.99950
3.3	.99952	.99953	.99955	.99957	.99958	.99960	.99961	.99962	.99964	.99965
3.4	.99966	.99968	.99969	.99970	.99971	.99972	.99973	.99974	.99975	.99976
3.5	.99977	.99978	.99978	.99979	.99980	.99981	.99981	.99982	.99983	.99983
3.6	.99984	.99985	.99985	.99986	.99986	.99987	.99987	.99988	.99988	.99989
3.7	.99989	.99990	.99990	.99990	.99991	.99991	.99992	.99992	.99992	.99992
3.8	.99993	.99993	.99993	.99994	.99994	.99994	.99994	.99995	.99995	.99995
3.9	.99995	.99995	.99996	.99996	.99996	.99996	.99996	.99996	.99997	.99997

Bibliography

Agresti, A. (2017). Statistical methods for the social sciences. Pearson.

Baruah, H. K., & Saikia, S. (2015). Statistical methods and applications. PHI Learning Pvt. Ltd.

Bhat, B. R., & Gupta, A. K. (2013). Elements of applied stochastic processes. John Wiley & Sons.

Chatterjee, S., & Hadi, A. S. (2015). Regression analysis by example (5th ed.). John Wiley & Sons.

Das, M. N., & Babu, G. J. (2015). Statistics for management. Pearson India.

Dash, G. S., & Dash, M. (2016). An introduction to statistics with applications in communication and engineering. Springer India.

De Veaux, R. D., Velleman, P. F., & Bock, D. E. (2016). Intro stats (5th ed.). Pearson.

Desai, J. P. (2016). An introduction to statistics (3rd ed.). Himalaya Publishing House.

Freedman, D., Pisani, R., & Purves, R. (2007). Statistics (4th ed.). W. W. Norton & Company.

Groebner, D. F., Shannon, P. W., Fry, P. C., & Smith, K. D. (2018). Business statistics (10th ed.). Pearson.

Gupta, S. C., & Kapoor, V. K. (2019). Fundamentals of mathematical statistics. Sultan Chand & Sons.

Jagan, K., & Singh, R. K. (2013). Applied statistics for engineers and scientists. Narosa Publishing House.

Joshi, R. M. (2018). Statistics for economics. Oxford University Press.

Jothi, V., & Narayanan, A. (2014). Probability, statistics, and queuing theory with computer science applications. PHI Learning Pvt. Ltd.

Khanna, V. K., & Srivastava, S. (2016). Fundamentals of applied statistics. Sultan Chand & Sons.

Kulkarni, V. G. (2016). Modeling and analysis of stochastic systems. Chapman and Hall/CRC.

Lahiri, S. N. (2017). Probability and statistics for engineers and scientists. Prentice Hall India.

Mahajan, R. (2018). Applied statistics for engineers and scientists. Springer India.

Misra, S. P., & Yadav, R. K. (2018). Statistical methods. Himalaya Publishing House.

Moore, D. S., McCabe, G. P., & Craig, B. A. (2017). Introduction to the practice of statistics (9th ed.). W. H. Freeman.

Pillai, R. S., & Bagavathi, A. (2017). Probability and random processes. McGraw Hill Education.

Rao, C. R. (2008). Linear statistical inference and its applications (2nd ed.). John Wiley & Sons.

Rao, M. C. (2017). Statistical methods: An introduction. Prentice Hall India.

Rice, J. A. (2006). Mathematical statistics and data analysis (3rd ed.). Cengage Learning.

Shivakumar, P. N., & Sathish, R. (2016). Business statistics. Vikas Publishing House.

Subba Rao, T., & Padma Raju, A. S. (2014). Statistical methods for forecasting. Wiley India Pvt Ltd.

Index

A

Aggregate method 198, 218,
Alternative hypothesis 157, 283, 584, 285, 286, 291, 292, 293, 296, 297, 298, 300
Arithmetic mean 80, 81, 82, 83, 84, 86, 88, 91, 92, 110, 125, 127, 132, 140, 155, 204, 206, 221, 232

B

Bar chart 27, 31, 33, 53, 65, 67, 68, 70, 78
Bias 6, 41, 44, 48, 56, 58, 131, 212, 216, 230, 232, 236, 240, 243, 244, 245, 256, 258, 259, 261, 268, 271, 273, 274, 278
Binomial distribution 303, 305, 306, 307, 308, 310, 319, 320, 321
Boxplot 8

C

Categorical rating scale 40
Categorical variable 9, 10, 11, 71
Central limit theorem 314, 315
Central tendency 6, 19, 30, 33, 35, 64, 79, 80, 81, 87, 88, 94, 95, 98, 103, 111, 112, 131, 137
Cluster Sampling 364, 365, 366, 267, 277
Coefficient of determination 160, 161, 191
Confidence limits 279, 290, 291, 300
Confounding variable 16, 17, 161
Consumer price index 24, 196, 216, 229, 232
Continuous probability distribution 303, 313, 320
Continuous variable 11, 39, 73
Control variable,16, 19
Convenience sampling 254, 269, 270, 271, 277
Correlation 22, 23, 43, 44, 46, 54, 141, 142, 143, 144, 145, 146, 147, 148, 149, 150, 151, 153, 154, 155, 156, 157, 158, 160, 161, 162, 163, 164, 165, 166, 167, 168, 169, 171, 179, 180, 181, 183, 184, 186, 192, 193, 286
Cost of living index 217, 218, 219, 232
Critical region 295, 296, 297, 298
Cross-tabulation 30
Cumulative frequency 33, 62, 63, 65, 74, 75, 76, 77, 78, 96, 97, 104, 106, 109

D

Data analysis 2, 3, 5, 6, 21, 22, 23, 26, 43, 45, 47, 49, 73, 79, 80, 103, 105, 110, 113, 237, 239, 276
Data collection 2, 3, 6, 12, 19, 41, 44, 49, 52, 53, 54, 55, 56, 57, 58, 77, 230, 231, 238, 239, 245, 247, 260, 265, 266, 267, 270, 271, 320
Deciles 79, 102, 105, 106, 110
Degrees of freedom 279, 289, 290, 300
Dependent variable 13, 14, 15, 16, 17, 161, 164, 165, 169, 171, 172, 175, 177, 191
Descriptive statistics 6, 8, 19, 27

Discrete probability distribution 303, 305, 319, 320
Discrete variable 11
Disproportional stratified sampling 261

F

Frequency distribution 7, 30, 33, 58, 59, 60, 61, 62 73, 74, 76, 77, 78, 99, 114, 115, 132, 133, 137, 158
Frequency polygon 7, 65, 74, 75, 76, 78

G

Geometric mean 80, 86, 87, 88, 89, 90, 91, 92, 110, 179, 203, 204, 205, 206, 221, 222, 224
Graphical presentation 63, 65
Graphical representation 3, 7, 19, 27, 33, 35, 62, 63, 76, 128, 149
Grouped data 60, 61, 62, 82, 104, 106, 108, 109, 154

H

Harmonic mean 80, 87, 91, 92, 93, 110
Histogram 7, 19, 27. 35, 53, 59, 65, 73, 74, 76, 77, 78, 99, 100
Human development index 223, 225, 229
Hypothesis testing 19, 23, 27, 121, 239, 240, 279, 280, 281, 282, 283, 285, 286, 289, 290, 291, 292, 293, 295, 296, 297, 298, 299, 300

I

Independent variable 13, 14, 15, 16, 17, 161, 162, 164, 165, 169, 171, 172, 173, 174, 177, 189, 190, 191, 280
Index number 193, 195, 196, 197, 198, 199, 203, 206, 207, 208, 209, 210, 211, 213, 214, 215, 216, 218, 219, 221, 223, 225, 226, 227, 228, 229, 230, 231, 232, 233
Industrial production index 221, 222
Inferential statistics 6, 8, 19
Interval 10, 11, 13, 19, 21, 22, 26, 27, 28, 31, 32, 33, 34, 35, 36, 37, 38, 39, 41, 45, 48, 51, 52, 58, 59, 60, 61, 62, 73, 74, 76, 80, 82, 83, 98, 101, 121, 122, 247, 251, 252, 262, 263, 264, 274, 290, 291, 303, 310, 320

K

Kurtosis 6, 8, 111, 131, 137, 138, 139, 140

L

Levels of significance 285
Likert scale 32, 41
Line chart 35, 53, 65, 66, 77,
Lorenz curve128, 129, 130, 140

M

Mean 2, 5, 65, 7, 8, 10, 19, 27, 32, 33, 34, 35, 36, 39, 46, 73, 77, 79, 80, 81, 82, 83, 84, 86, 87, 88, 89, 90, 91, 92, 93, 94, 95, 97, 98, 101, 102, 103, 107, 110, 112, 113, 114, 117, 118, 119, 120, 121, 122, 124, 125, 126, 127, 128, 129, 132, 133, 134, 135, 137, 140, 143, 144, 145, 147, 149, 155, 157, 168, 169, 174, 176, 179, 181, 182, 183, 185, 186, 189, 191, 193, 199, 203, 204, 205, 206, 216, 221, 222, 223, 224, 236, 250, 251, 263, 277, 280, 281, 282, 283, 284, 286, 288, 289, 290, 292, 293, 294, 295, 296, 297, 298, 300, 301, 305, 309, 310, 311, 312, 313, 314, 315, 316, 317, 319, 320, 321
Mean deviation 117, 118, 119, 120, 128, 140
Measurement scale 9, 21, 26, 30, 38
Measures of dispersion 19, 111, 112, 113, 114, 139
Median 6, 7, 19, 27, 33, 36, 62, 73, 75, 76, 79, 80, 94, 95, 96, 97, 98, 101,

102, 103, 110, 113, 116, 117, 118, 119, 120, 134, 135, 137, 140, 154, 313, 314, 321
Mediating variable 14, 15
Mode 6, 19, 27, 28, 30, 36, 73, 79, 80, 94, 95, 97, 98, 99, 100, 101, 102, 110, 134, 137, 313, 314,
Moderating variable 15
Moments 111, 131, 132, 134, 135, 138, 139
Multiple regression 165, 171, 172, 182, 192

N

Nominal scale 27, 28, 29, 30, 31, 32, 38, 39, 52
Non-Probability sampling 235, 254, 255, 269, 271, 273, 276, 277, 278
Non-Sampling error 235, 274, 275, 277
Normal distribution 8, 87, 157, 238, 294, 303, 313, 314, 315, 316, 317, 318, 319, 320,
Null hypothesis 157, 239, 280, 283, 284, 285, 286, 287, 288, 289, 291, 292, 293, 295, 296, 297, 298, 300
Number of trials 305, 310, 320
Numerical variable 10, 11

O

Ogive 62, 65, 75, 76, 78
Ordinal scale 26, 27, 28, 31, 32, 33, 34, 38,
Outliers 6, 8, 45, 47, 59, 63, 73, 80, 81, 87, 92, 94, 95, 98, 103, 108, 112, 114, 115, 117, 131, 149, 175

P

Panel data 13, 47, 48
Partition values 79, 102, 110
Percentile 7, 33, 75, 76, 79, 102, 103, 105, 107, 108, 109, 110, 295
Pie chart 19, 31, 53, 65, 71, 72, 77, 78
Poisson distribution 303, 310, 311, 312, 313, 320, 321
Price index number 198, 199, 206, 208, 211, 215, 216, 221, 227
Probability sampling 235, 247, 249, 253, 254, 255, 256, 259, 262, 268, 269, 271, 273, 276, 277, 278
Proportional stratified sampling 261
Purposive or Judgmental Sampling 271

Q

Quality control 3, 4, 5, 23, 112, 113, 117, 121, 305
Quantity index number 198, 211, 215, 216, 221
Quartile 7, 8, 75, 76, 79, 102, 103, 104, 105, 107, 108, 110, 111, 115, 116, 128, 134, 140
Quartile deviation 115, 116, 128, 140
Quasi-Random sampling 274
Quota sampling 273, 274, 277

R

Range 1, 4, 6, 7, 11, 19, 22, 25, 32, 36, 41, 47, 57, 59, 61, 65, 73, 76, 87, 95, 98, 103, 111, 112, 113, 114, 115, 121, 128, 130, 139, 1441, 142, 144, 147, 155, 160, 161, 164, 168, 191, 217, 238, 250, 251, 258, 259, 263, 290, 291, 303, 304, 318
Range variance 73, 111, 139
Ranking scale 39
Rating scale 39, 40, 41
Regression 3, 19, 27, 142, 144, 153, 161, 165, 168, 171, 172, 173, 174, 175, 176, 177, 178, 179, 180, 181, 182, 183, 184, 185, 186, 187, 188, 189, 190, 191, 192, 193
Regression line 171, 173, 175, 179, 180, 182, 183, 184, 185, 186, 192, 193
Relative frequency distribution 59, 61, 77
Relative method 206, 208

S

Sample 3, 5, 6, 8, 9, 13, 17, 19, 46, 55, 56, 58, 74, 95, 98, 157, 160, 220, 222, 235, 236, 237, 238, 240, 241, 243, 244, 245, 246, 247, 248, 249, 250, 251, 252, 253, 254, 255, 256, 257, 258, 259, 260, 261, 262, 263, 264, 265, 266, 267, 268, 269, 270, 271, 272, 273, 274, 276, 277, 278, 279, 281, 284, 286, 288, 289, 290, 291, 293, 294, 295, 296, 300, 312, 314
Sampling error 5, 235, 236, 238, 247, 270, 274, 275, 277,
Sampling technique 218, 235, 237, 249, 253, 256, 264, 265, 269, 273, 274, 276, 277,
Scatter plot 27, 25, 142, 149, 150, 172, 173, 174, 175
Semi-Interquartile range 115
Simple random sampling 254, 256, 257, 258, 259, 260, 262, 263, 264, 265, 266, 277
Skewness 6, 8, 87, 111, 131, 133, 134, 135, 136, 137, 138, 139, 140
Snowball sampling 273, 277
Stacked bar chart 68, 78
Standard deviation 6, 7, 19, 27, 33, 35, 36, 39, 88, 111, 112, 113, 114, 115, 117, 120, 121, 122, 123, 125, 126, 128, 134, 139, 140, 147, 155, 180, 181, 193, 248, 251, 252, 278, 294, 296, 300, 301, 305, 314, 315, 316, 317, 318, 319, 321
Stapel scale 40
Statistical techniques 2, 6, 23, 27, 230, 240, 247, 250
Statistics 1, 2, 3, 4, 5, 6, 7, 8, 9, 10, 11, 17, 18, 19, 21, 22, 23, 27, 57, 63, 74, 95, 97, 102, 111, 113, 131, 139, 186, 195, 196, 197, 231, 235, 239, 246, 248, 265, 281, 282, 289, 290, 291, 305, 314
Stratified random sampling 259, 260, 261
Systematic sampling 249, 262, 263, 264, 271, 277

T

Tabular presentation 63
Textual presentation 63
Time series data 45, 47, 49, 65
Type I error 283, 285, 287, 288, 289, 290, 293
Type II error 279, 283, 288, 289, 293, 300

U

Ungroup data 60

W

Wholesale price index 220, 221, 227, 229